In Praise of Difference
Genetics and Human Affairs

IN PRAISE OF DIFFERENCE

Genetics and Human Affairs

ALBERT JACQUARD

Translated by Margaret M. Moriarty

New York COLUMBIA UNIVERSITY PRESS *1984*

Library of Congress Cataloging in Publication Data
Jacquard, Albert.
In praise of difference.

Translation of: Eloge de la différence.
Bibliography: p.
Includes index.
1. Human genetics.   I. Title.
QH431.J27713 1984   573.2′1   84-7002
ISBN 0-231-05482-3 (alk. paper)

Columbia University Press
New York   Guildford, Surrey
© 1978 Editions du Seuil
English translation © 1984 Columbia University Press
All rights reserved

Printed in the United States of America

*Clothbound editions of Columbia University Press Books are
Smyth-sewn and printed on permanent and durable acid-free paper*

CONTENTS

PREFACE

The Improvement of the Human Race: An Ancient Objective

IT IS an essential property of Man to transform the world around him; his nature is to live among his own artifacts. He manipulates his environment, including various plant and animal species, to his own advantage. Human endeavor, based on a growing knowledge of both the inanimate and the living world, has become increasingly effective. Why not use this new power to achieve the most tempting objective of all: the improvement of the human species itself?

This idea is a very old one. Man has long considered himself responsible for his moral or spiritual transformation, as a means of achieving a better society. In addition, he has also been interested in his physical and biological welfare. Long ago, the Egyptians, the Hebrews, and the Greeks were concerned about such things as the preservation of their "race" from eventual degeneration. They hoped to improve at least part of their group and to arrive at a new Man, endowed with superior faculties. In our own century, new hopes have been kindled due to a combination of factors: the abandonment in the nineteenth century of static theories according to which each species is a specific and definitive creation of God; the discovery of the process of transmission of biological characteristics between generations; and the growing recognition of the link between the genetic heritage of an individual and the traits he manifests. Is it possible that we will soon be able to mold the human species?

Before dealing with the vague hopes and fears aroused by this prospect, it is important to define what we already know and what we wish to achieve; what, exactly, is the problem?

In order to give a valid reply, it is necessary to have a good understanding of recent advances in scientific knowledge. Unfortunately, the terminology used to transmit this knowledge is often ambiguous and confusing. Words drawn from everyday speech are sometimes used by specialists to designate many different concepts. Instead of facilitating the exchange of ideas, the use of such words hinders communication. In this little book, at risk of seeming to state the obvious, many passages are devoted to defining "words." For example, what is the exact meaning of the words: biological heritage, race, intelligence, improvement, consanguinity, etc.?

Even the word "genetics" leads to confusion. It evokes a scientific discipline that appears to be crowned with success. How can we help but marvel at how much has been discovered in the short time since its foundation at the beginning of this century? In relatively quick succession, many major break-throughs have been made. The structure of the genetic material (the famous double helix of deoxyribonucleic acid, DNA, which contains the basic genetic information in coded form), the mechanisms of protein production (the genetic "code") and of genetic regulation (the "operon") have all been elucidated. Seemingly insoluble mysteries have been explained. They are now represented in school textbooks by perfectly clear scientific models. The subdiscipline of molecular and cellular genetics is constantly adding to our knowledge of the functioning of living things. In a completely different area, that of plant and animal breeding, progress has been equally remarkable. New methods of selection, of crossing, and of hybridization have led to increases in the yields of milk, meat, and corn that would have been considered impossible a century ago. Quantitative genetics, the branch of genetics involved in the development of these methods, has thus been marvelously successful in that it has made it possible for humans to *act* on nonhuman species for their own benefit.

Those interested in transforming our own species would be obliged, not merely to alter a single individual with a limited life span, but to irreversibly modify our common biological heritage. The branch of genetics concerned would be neither of those already mentioned (molecular and cellular, and quantitative) but population genetics. As a scientific discipline, population genetics is still emerging amid a good deal of internal controversy.

This branch of genetics, less well known because it is more abstract and mathematical, attempts to outline, at the level of whole populations, the consequences of our post-Mendelian vision of the process by which traits are transmitted from parents to offspring. Its essential aim is to explain the evolution of living things, to understand the mechanisms whereby the living world could have materialized, starting from the few molecules of DNA that were scattered in the primordial "soup."

If man wishes to influence the course of nature and to play the game of life by his own rules rather than by those set by the traditional players, whether these be chance, necessity, the gods or God, he must face up to the extent of his knowledge, or rather of his ignorance.

The only aim of this book is to take stock of the principal problems, provisionally, of course. The reader will find more doubts than certainties, more questions than answers. However, freeing oneself from the illusion of understanding and getting rid of accepted ideas is a first step toward knowledge.

In Praise of Difference
Genetics and Human Affairs

ONE

The Elementary Process: Parent Begets Child

ANY DISCOURSE pertaining to genetics has as its basic reference point the obvious fact that there is a certain similarity
between children and parents. Throughout the living world, the
transmission of particular traits is part of the transmission of life
itself. What is the mechanism that governs this transmission? The
answer to this question seems to defy common sense. A rereading
of some classical texts illustrates the difficulty of finding a reasonable solution to this problem, even when it is posed in simple terms.

Some Pre-Mendelian Notions

Some authors confined themselves to observation and description, without suggesting an explanation. For instance, Ambroise
Paré wrote, at the end of the seventeenth century:

Children resemble their father and mother not only physically (in size, in
weight, in posture and in build) and in their manner of speaking and walk-
ing: they are also subject to the same diseases as those to which their father
and mother are subject, called hereditary diseases. This is seen to be true
of people afflicted with leprosy, gout, epilepsy, kidney stone, melancholia
and asthma. Thus, when one of these people engenders a child, it is likely
to be afflicted with the same disease. However, experience shows that this

does not always happen. It is prevented by the good quality of the woman's seed and the favourable conditions in her womb which correct the imbalance in the male seed just as the man's seed can correct the woman's.

Similarly, Montaigne expressed dismay at being afflicted with kidney stones at the same age as his father. How, he wondered, had the latter transmitted this disorder to him?

> How did this slight bit of his substance, with which he made me, bear so great an impression of it for its share? And moreover, how did it remain so concealed that I began to feel it forty-five years later, the only one to this hour out of so many brothers and sisters, and all of the same mother? If anyone will enlighten me about this process, I will believe him about as many other miracles as he wants; provided he does not palm off on me some explanation much more difficult and fantastic than the thing itself.[1]

What an apt warning to pedants whose explanations are more complex and mysterious than the question asked!

Others dreamed up theories which seem very strange to us and, indeed, it is difficult to understand how they could ever have been proposed as rigorous scientific theories. For example, Buffon thought that the male and female "seminal fluids" contained particles emanating from all parts of the body, which fell miraculously into place to constitute a child.

> I therefore believe that the male seminal fluid, which is spread on the vagina, and the corresponding female fluid which is spread in the uterus are both equally active substances and equally loaded with organic molecules, endowed with engendering properties.
>
> I think that, when the two seminal fluids mix, the activity of the organic molecules in each of the fluids is, in some manner, stabilized by the counterbalancing action of the other, in such a way that each organic molecule becomes immobile at an appropriate location, this being none other than that which it occupied previously in the body. Thus, all the molecules which came originally from the spinal cord will be relocated in a comparable way, in relation both to the structure and to the position of the vertebrae.

1. Donald M. Frame, tr., *Montaigne's Essays and Selected Writings: A Bilingual Edition* (New York: St. Martin's Press, 1963).

Darwin's attempt at explaining the transmission of traits was scarcely more convincing than Buffon's. He proposed that the various characteristics and functions of each cell in an organism are dependent on one or more very small particles called gemmules. Gemmules from both father and mother meet in each of the embryo's cells, which are thus endowed with characteristics that are intermediate between those of the corresponding maternal and paternal cells.

Let us forget what we may have learned and try to imagine how one individual might be generated from two others. This everyday event seems so inexplicable that one's first thought is to suggest that only one of the two parents plays a real part. Such was the point of view adopted by the "spermatists," according to whom each sperm contains a fully formed baby who merely grows bigger in the mother's womb.

This theory was developed soon after the first observations of sperm through a microscope; these had revealed the presence of tiny living particles, called "homunculi." It was readily accepted because it gave an easy answer to certain problems, for instance, that of original sin. Some Christians were unwilling to accept responsibility for a sin that they had not committed. The spermatists' theory appealed to them because it included the notion of "boxes within boxes": the baby in the sperm has itself testicles inside which are sperm each containing a baby which itself . . . All generations, past and future, are thus nested in each other, like a series of Russian dolls, from Adam down to the end of the world. We were therefore present in Adam's body when he rebelled against God; it is right that we are punished! Thus, a biological theory develops and is generally accepted only to the extent that it corresponds to the preoccupations of the time. It therefore runs the risk of being deflected from its real goal and of being used as a justification for arguments outside of it; we will see some contemporary examples of this later on.

The "spermatists" were challenged by the "ovists." These claimed that, on the contrary, the baby is prefabricated in the mother's egg and that the sperm merely triggers the developmental process, without contributing anything essential.

The problem with both of these theories is that the offspring receives its biological heritage from only one of the two parents and, therefore, has no reason to resemble the other; this prediction is clearly contradicted by the evidence.

Buffon's fluids theory and Darwin's gemmules theory, which was fairly universally accepted at the end of the nineteenth century, both admit that the two parents participate equally in the production of their child. However, this idea also leads to an insurmountable paradox. It suggests that each of the child's traits represents the average between those of its parents. Within a population taken as a whole, the variability of traits among individuals should therefore diminish with each new generation; rapidly, all individuals should become, if not identical, at least, very similar. Again, this does not correspond to our observations.

This "variance paradox" could not be overcome until the discovery of entirely new concepts. The "model" allowing us to understand the mechanism of sexual reproduction was first proposed by Gregor Mendel, a monk in a monastery in Brno, Czechoslovakia. As early as 1865, he had published a new explanation for the transmission of traits. However, on account of their very novelty, his ideas were ignored. Not until 1900, in a more receptive intellectual environment, could this model which has since become the foundation of the science of genetics finally be understood, accepted, and developed.

Mendel's Contribution

Mendel's extraordinary genius lies in not having sidestepped the essential difficulty encountered during his hybridization experiments: the disappearance and subsequent reappearance of various traits from one generation to the next.

Let us imagine the following "experiment." We populate an island with women chosen from a population where everyone, for numerous generations, has been of the Rhesus positive blood group, and with men from a population where everyone belongs to the Rhesus negative group. One notices that their children all belong to the positive group; the negative trait has disappeared completely from this first generation. In the next generation, however, this trait reappears and is manifested by about a quarter of the grandchildren.

This remarkable result is observed every time we repeat the experiment, each time with the same frequency of one quarter.

Since it reappears, the Rhesus negative trait must have been present in some of the children of the first generation; but under what form? Why did it not manifest itself?

Mendel, observing generations of peas rather than of people (which made the experiments simpler without in the least altering their basic significance), had noted precisely the same phenomenon as that which has just been described. His ingenious idea was to suggest that each trait under study (in our imaginary example, the Rhesus system; in his own experiments, the green versus yellow color of the cotyledons and the round versus wrinkled surface of the seeds) is governed not by *one* hereditary factor, but by *two* factors, one of which comes from the father, the other from the mother. These two factors function in concert; the trait under observation results from the activity of both; but they remain unchanged throughout the entire life of the individual. They coexist, but they do not modify each other. When an individual procreates, he transmits one of these factors, chosen at random, to the child.

In Mendel's case, this could be no more than a hypothesis or, as we say today, a model. The progress achieved in our knowledge of cells, their nuclei and their chromosomes, has shown that this theory corresponds to reality in every respect. The "factors" mentioned by Mendel are what we now call "genes," and are series of chemical molecules situated at specific locations on the chromosomes. Their mode of action and of transmission from parents to offspring is consistent with Mendel's theory.

Each cell belonging to a hypothetical human individual, X (and his body is made up of hundreds of billions of them), has a nucleus containing 23 pairs of filaments called chromosomes. These 46 filaments represent an exact copy of the 23 chromosomes supplied by the paternal sperm and of the 23 chromosomes supplied by the maternal egg. The various processes necessary for the development and functioning of the organism are defined and regulated by information inscribed in coded form (the famous genetic code) on the chromosomes. Each of X's cells, whether belonging to his liver or to his brain, knows the amazing secret which allows for the construction of X in his entirety, starting from a single cell. Besides being a liver or nerve cell with a set of specific functions, it "knows" that it belongs to X and is recognized as such by its neighboring cells.

The sex cells, however, are exceptions: the sperm made by X (if a man), the eggs (if a woman), contain a series of only 23 chromosomes, one of each pair. They therefore contain only half of the genetic information that X received at conception. The process of generating eggs and sperm is such that this half is drawn equally from material contributed by X's father and X's mother. It is clear that this biological mechanism affects the transmission of traits in a manner that coincides exactly with the Mendelian model.

Such a vision of the hereditary process profoundly changes our spontaneous "commonsense" ideas on the subject, but it is far from easy for this change to become part of our consciousness. This is shown by the contradiction that often exists between the usual connotations of the words that we use to describe heredity and the precise meaning that must now be ascribed to them.

Words and Their Meanings

Each word has a precise meaning, ascribed specifically to it and to no other. At the same time, it connotes a particular world view, which is revealed partly by its etymology. It is remarkable that the statements we make about procreation often harbour a contradiction between their precise meaning and this world view. For example, the statement "an individual reproduces himself" is doubly contradictory.

The word *individual* suggests *indivisibility*. An individual cannot be reduced to his constituent parts without being destroyed. In other words, he cannot be divided. However, in the course of reproduction, such a division must, of necessity, take place. More precisely, each sperm or each egg receives a copy of half the information that had initially been transmitted to this individual by his parents at conception, and based on which he was gradually constructed.

It is necessary to understand thoroughly that this mechanism is the very opposite to that accepted prior to Mendel, for instance by Darwin. The latter suggested that each parent transmits all his biological information to the child, in accordance with the concept of indivisibility; the two stores of information, one from the father and one from the mother, mix together to form an average, just as white and red liquids mix and become pink. Mendel suggested, on the contrary, that each parent contributes only half of the information which he carries; in the child, these two halves are juxtaposed, without being mixed, to constitute a complete whole. Moreover, this whole, which is really a collection of different pieces of information, is entirely new, differing as much from one parent as from the other.

"Reproduction" does not, therefore, take place. This word implies the making of an image as close as possible to the original. Such is indeed the case with bacteria, which are capable of self-duplication by producing an exact copy of themselves, and with nonsexual species in general. However, the invention of sexuality, that is of a mechanism necessitating the collaboration of two beings for the making of a third, eliminated this capacity for reproduction. A sexual being cannot reproduce itself. Since the child is not a reproduction of anyone, it is a definitively unique creation. This uniqueness is a result of the amazing number of different children that could be produced by any one couple. Let us imagine that, for a given trait, for example, the Rhesus blood group, the father and mother each carry two different genes, a and b; their children can receive either two a genes or two b genes, or an a gene and a b gene. For each trait, 3 combinations are therefore possible; for two traits together, $3^2 = 9$ combinations; for n traits, 3^n combinations. This last number becomes astronomical as soon as n is larger

than 30 or so. Thus, for a total of 200 traits, the number becomes 3^{200},[2] which is practically infinite since it gives rise to a 94 digit number, billions of times greater than the total number of atoms in our universe, including the farthest galaxies.

This possibility for diversity is the specific contribution of sexual reproduction. There may be but a single realized event, but the possibilities are infinitely great. How is this reality chosen? To explain this, it is necessary to introduce a concept that is poorly described by the time-worn word associated with it: chance.

Chance

This term is used in so many different ways that it no longer has a precise meaning. We will use it here only after eliminating the metaphysical connotations which would make of chance an all-powerful and unreachable god, with an objective existence and a will of its own. Chance, as defined by scientists, is related to knowledge and prediction. When confronted with a particular aspect of reality, one tries to understand how the phenomena observed are interrelated; one then uses this understanding to predict future phenomena. Thus, an understanding of the movement of stars and of the attraction between masses allows one to precisely predict the future movements of the stars, or the eclipses of the sun.

However, very often, this knowledge is not complete enough to make prediction possible. For instance, when a dice is thrown, we cannot predict the outcome because the phenomena involved are too complex and too poorly known. We therefore say that the outcome is a matter of chance. We are, however, capable of imagining that our knowledge will one day be more complete and that, with a better understanding of the characteristics of the dice and of the strength of the initial throw and of the resistance of the air . . . , we will be able to predict the outcome with certainty. Chance will have given way to determinism.

2. That is, 3 multiplied by itself 200 times.

Such a solution does not, however, seem likely in the case of the formation of a sperm or an egg, if we wish to predict, for each unit trait, which gene copy will be transmitted. No doubt, events occur in a deterministic fashion at the molecular level, but the number of possible outcomes is so great that there is scarcely any hope of arriving at a sufficiently accurate knowledge of the phenomenon. The facts, as we know them, suggest that, in this case, we must, for once and for all, rely on chance to explain reality. Our only means of pinning down the process better is to precisely calculate the probability of each result. The essence of Mendel's laws is the claim that each gene, whether paternal or maternal, has an equal probability, one-half, of being transmitted.

Thus, the transmission of the genetic heritage seems to depend on an immense number of lotteries that, in the case of each trait, determine which of the two genes present is to be chosen. Sexual reproduction puts chance at the heart of the phenomenon, "chance" in the sense of *the ensemble of factors which intervene or appear to intervene in a process, but whose precise mode of action we may never understand.*

Therefore, in order to understand the transmission of life, it is not enough to assume the existence of the two principal players, the father and the mother. They provide copies of their collections of genes; but the choice of half of one of these collections along with half of the other, to make up a complete collection, is determined by chance.

"Genotype" and "Phenotype"

We now see more clearly why it is necessary to distinguish between two aspects of each living being. On the one hand, there is the individual whom we see, single, indivisible, living out his unique experience of development, aging, and eventual death. On the other hand, there is the collection of genes which he possesses, genes with a multitude of functions, derived from two immediate sources, the father and the mother, and which are capable of making an unlimited number of copies of themselves, unchangeable, impervious to the onslaught of time, quasi-eternal, since they will still be exactly the same when they reappear in the children or grandchildren long after the death of the parent.

This duality is fundamental. The failure to recognize it is the basis for most misunderstandings regarding the transmission of traits. It is useful to set this duality down in words. Unfortunately, the terms at our disposition are very technical:

— the "phenotype" denotes an individual's *appearance* or, more precisely, the ensemble of those traits that can be measured or described. Some of these, for example the blood groups, are in fact not at all apparent and can be identified only after complex investigation.
— the "genotype" denotes the *collection of genes* which an individual receives at conception.

The study of the transmission of traits involves the description of the exact mode of interaction between genotype and phenotype while also, of course, taking the role of the environment into account. This interaction is necessarily complex and all simplistic explanations should be subjected to rigorous analysis. Likewise, it is necessary to be especially wary of numerical conclusions, which, buttressed by lengthy argument and laborious calculations, can give a false impression of a clear understanding of the problem. The only serious scientific approach is one that respects the reality: if that reality is complex, it cannot but be betrayed by simplistic presentation.

When we consider a succession of individuals over many generations, within a single pedigree, it is necessary to realize that a "trait" manifested by any one individual depends:

— on the genes which he has received from his father and mother (and half of which he will transmit to his own offspring),
— the effects on this trait of various "environmental" factors, including both the physical environment (heat, humidity, nutrition . . .) and the social environment (family, school, society . . .).

The resemblance between parents and offspring, implied by the saying "like father, like son," is therefore the result of numerous factors which are difficult to explore; and the saying itself is, in fact, only partly accurate.

In this field of study, the only objective reality is the transmission by halves of the genetic complement. The unchanging nature of this genetic heritage justifies the maxim that there is no "inheritance of acquired characteristics." Indeed, the changes which an individual undergoes as a result of his personal experiences cannot in any way change the structure of his genes. With the exception of mutations (extremely rare events), he transmits his genes exactly as he received them, without any trace of the events which he experienced during the course of his lifetime.

However, the above maxim corresponds to the reality only in this limited sense. When we consider the total picture, we see that the trait manifested by the child depends not only on the inalterable genes transmitted by his parents, but also on the influence exercised by those parents during his upbringing, which, in the case of our own species, requires an especially long period of time. Their influence will depend, naturally, on the sum total of their own experiences, a fact which provides an undeniable role for "acquired characteristics."

Likewise, it would be wrong to equate the biologist's distinction between "genotype" and "phenotype" with the "nature-nurture" duality of the philosopher. It is true that the genetic complement contains all the information necessary for the development and functioning of the organism. Another similar organism can be produced from this single collection of genetic instructions, as can be shown simply by the propagation of plants from cuttings: an entire tree, biologically identical to the original, can be obtained from a single branch because that branch contained within its cells the information for the whole tree. However, in the case of a human being, it is difficult to explain him in terms of the genetic rules which governed his development; the many other events which have shaped him, and which have contributed to his phenotype, are as much a part of his nature as are the genes which initiated the process. We will see, with regard to the relationship between intelligence and genetics, how dangerous it is to accept such analogies uncritically.

The analogy that is undoubtedly the most satisfactory used to date is a musical one: the genotype is the score, the phenotype is the symphony which we hear and which is influenced by the personality of the conductor and by the ability of the performers. In the case of genetics, we have scarcely any access, with our present limited knowledge, to the score. We have to be content with observing the interpretation which is rendered by the environment, that is, by the ensemble of events which make up the lifetime experience of each person.

Let Us Make a Sketch

The distinction we have just made between genotype and phenotype is fundamental. Let us make a sketch to show it more clearly (figure 1). The lower level refers to the genotypes: three individuals, a father F, a mother M, and their child C, are each symbolized there by a circle. Inside the circles, two lines represent the two collections of genes which constitute the genetic complement of each of them. An arrow from M to C and from F to C indicates the transmission of half of this complement; these arrows originate from arcs of circles, thus suggesting the fact that the halves transmitted to C are chosen randomly.

This description of genotypes is remarkably simple and it conveys the essence of the "Mendelian model." The situation is quite different, however, in the upper part of the sketch which deals with phenotypes. Here we again have the same three individuals, father, mother, and child, now represented by silhouettes, because we are dealing now with the traits they actually manifest and not with the genes hidden in the nuclei of their cells.

Columns link the circle symbolizing each individual's genotype with the silhouette symbolizing his phenotype; these columns are meant to suggest not the transmission of material things but a dependent relationship. The phenotype realizes itself little by little based on the information contained in the genotype. As the need arises, chemical substances, mostly proteins, are made by the organism following procedures and methods of assembly that are precisely specified by the genotype: we can say that the latter

Figure 1

"governs" the phenotype. Thus, phenotype depends on genotype, but the reverse is not true. That is, regardless of what adventures an individual may experience during his lifetime, his genotype remains unchanged, "acquired traits" are not incorporated into the genes.

The upper part of the diagram is complicated by the fact that C's phenotype is influenced by the phenotypes of his parents and by the "environment" as a whole. The word "environment," which is deliberately imprecise, covers such things as the nourishment absorbed by C, the radiation he was exposed to, the shocks he endured, the affection with which he was surrounded, the teaching, of all kinds, which he was given: in short, all the physical and moral influences which shaped the individual in the course of its development from the embryo.

Accidental Events: Mutations

The beautiful simplicity of the lower part of our sketch does not in fact always correspond to the reality. Occasionally, the child C receives genes which were present neither in the genetic complement of his father nor in that of his mother. More precisely, an accident occurred at some stage during the transmission, and one particular gene with a given well-defined effect was transformed to become a gene with a different effect; in other words, a "mutation" occurred.

Countless studies, mostly of microorganisms or of the well-known drosophila flies, have made it possible to demonstrate the influence of certain factors (for instance, various types of radiation) on the frequency of these mutations, but the estimates of these frequencies are quite imprecise. The frequency of mutation seems to be in the region of 1 per 100,000 or 1 per million. It is, therefore, an extremely rare event; it is nevertheless quite important because it is the only means of genetic innovation for the species as a whole. The evolution of the living world, including the appearance of new species with new functions, necessitates changes in the genetic complement that can only be brought about by mutations. In addition to single mutations involving isolated genes, rearrangements involving entire sections of chromosomes are also seen to occur, and their role in the evolution of species seems to have been of the utmost importance.

Unit Traits and Complex Traits

We notice that the word "trait," which we have often used, also contains a trap.

When we are dealing with a blood-group system, for instance

the Rhesus system, the phenotype depends very directly on the genes which control it. In this case, a knowledge of the genotype allows us to predict the phenotype (without the opposite being true, since many different genotypes can give rise to a single phenotype). In the case of the Rhesus system, at least according to our simplified vision of it, the correspondence is direct. The genes belong to two categories, let us call them *R* and *r;* since each individual possesses two genes, three associations, or three genotypes, are possible: *RR, Rr* and *rr.* The first two correspond to the Rhesus positive phenotype, the third to the Rhesus negative phenotype. In other words, the gene *R* gives rise to the positive trait, whether it is present in double dose (as in *RR* individuals, called "homozygotes") or in single dose (as in *Rr* individuals, called "heterozygotes"). On the contrary, the gene *r* produces the negative trait only when it is present in double dose (in "homozygous" *rr* individuals). In this case, and as had been found by Mendel for the traits that he studied in peas, we say that the *R* gene is *dominant* while the *r* gene is *recessive.*

Such an unambiguous and direct determination of the phenotype by the genotype is encountered only in the case of certain traits, whose expression is strictly controlled by a single pair of genes. These traits are called unit or Mendelian traits (because Mendel's laws are directly applicable to them). In this category, we can place the many blood group systems, diseases caused by inborn errors of metabolism and some seemingly insignificant traits such as the ability to roll the tongue into the shape of a tube, or the ability to taste Phenyl-Thio-Carbamide, a synthetic chemical product. Each of us either has or has not the ability to roll our tongue, or to taste this product, depending on whether we have or have not the particular gene which underlies this ability.

Even in cases where this simple model, dealing with only one pair of genes, is applicable, one can still describe situations where there is a fundamental difference between the phenotype and the genotype. Let us imagine a defect caused by a recessive *d* gene (as is the *r* gene for the Rhesus system). Only homozygous individuals *dd* are affected; the others, whether *Dd* or *DD*, are healthy. In the island A, with 100 inhabitants, 40 are *Dd*, 60 are *DD;* in another island B, also with 100 inhabitants, the distribution is 10 *dd* and 90 *DD*. From a medical point of view, B island is reputed to be in poorer health since 10 percent of its inhabitants manifest the defect while no one on island A manifests it. To the geneticist, island A is, on the contrary, more severely affected since the frequency of the defective gene is 20 percent, while it is only 10 percent on island B. The two points of view are contradictory, but the facts are perfectly compatible. The two viewpoints focus on two different aspects of the situation: one on the obvious reality, the other on a deep, inaccessible reality, which is of critical importance to future generations. If we consider not just the present inhabitants of these islands, but their children in the future, we can calculate that the risk of having an affected child is 4 times greater on island A. This result is based on a general theory that is explained in the next chapter, and that deals, not with the genetics of individuals, but with that of whole populations.

In fact, for most "characteristics" in the usual sense of the word, this simple model, depicting a correspondence between a single pair of genes and the characteristic (or trait) under study, is not consistent with the observations. The mechanisms involved in the expression of a trait are numerous and intermingled: they depend on a multiplicity of genes as well as on the environment. The trait that one actually observes in an individual cannot be fully explained by those found in his parents.

In stressing this difficulty, we do not seek to minimize the importance of genetic research. All scientific research proceeds by analysis of the simplest cases; thus, physicists study "perfect" gases or "perfectly" elastic solids, while knowing quite well that such perfection cannot be found in nature. We merely wish to underline the precautions necessary before extrapolating the results of simple models to the complex problems of the real world. We will see in subsequent chapters that these precautions are rarely taken. Unfortunately, many statements that are couched in impressive scholarly language and bolstered with references to science are no more than high-sounding and confused nonsense, including everything from statements of the obvious to selfcontradiction.

TWO

The Collective Process: The Structure of Populations and the Succession of Generations

ALL OF THE discussion in the preceding chapter concerns individuals: a particular pair of parents produces a given child. But when we consider "Humanity," a race, or a nation, we imply a group of individuals; the object of our attention, at the level of "phenotypes" now becomes the society which those individuals comprise, and, at the level of "genotypes," the collective genetic heritage which they possess.

The object of study of the discipline which has developed since the beginning of this century under the name of Population Genetics is the examination of changes in this genetic legacy, or genetic store, called the "gene pool." In order to undertake such a study, one must first define several characteristics of the gene pool under consideration and then construct some hypotheses concerning the factors that will affect succeeding generations; in other words, one constructs theoretical "models" that may, or may not, be realistic.

Each individual inherits several hundred thousand pairs of genes, each pair capable of governing a particular elementary function, for example, the synthesis of an enzyme. Even if we were to limit ourselves to a few hundred such functions whose genetic basis has already been determined for the human species, we are still very far from being able to fully understand the genetic structure of a population of individuals. Usually, one is limited to estimating the frequencies of the various possible genotypes in the population under study.

For example, in the case of the Rhesus blood group system which was mentioned in the preceding chapter: if one simplifies the system appreciably, one could say that two alleles, R and r, are involved, such that in a population there are three genotypes, two homozygotes, RR and rr, and the heterozygote genotype Rr. But the number of alleles occupying a given chromosomal location, and thus affecting the same function, could be very great, sometimes several dozen. Let this number equal n. One would therefore encounter n homozygous genotypes and $n(n-1)/2$ heterozygotes. Thus, for the "ABO" blood group system, there are four alleles: A_1, A_2, B, and O and ten genotypes: the four homozygous genotypes A_1A_1, A_2A_2, BB and OO, and the $(4 \times 3)/2 = 6$ heterozygous genotypes A_1A_2, A_1B, A_1O, A_2B, A_2O and BO.

Knowing the frequencies of these genotypic combinations in a population allows us to define its "genetic structure." Of course, this information is usually unavailable. To simplify things, one can limit oneself to looking for the frequency of various alleles in the global genetic pool, without trying to calculate their associations as pairs in individuals. One can then compare two populations with regard to the frequency of particular alleles within them.

A great many such studies have been carried out, making it possible to draw maps of the world where the lines that usually join points at the same altitude (isoclinal lines) or with the same temperature (isotherms) are replaced by lines that join points where similar frequencies for a particular allele have been found. This "geographic hematology," as it was labelled by Professors Bernard and Ruffié (1966), has been extensively developed in recent years, leading to the publication in 1976 of an atlas (Mourant et al.) which describes, in over one thousand pages, the distribution on our globe of the genes corresponding to 67 genetic systems. We will return to the question of how to interpret these data which throw new light on the study of human "races" and may contribute to research on the great human migrations and on the way in which various countries were populated.

For the moment, we will concentrate on one rather surprising discovery, the first of note to be made by population geneticists: a knowledge of allelic frequencies allows one to calculate very precisely the frequencies of the various combinations of pairs of those alleles in genotypes.

A Well-Known Law: The Hardy-Weinberg Rule

It seems clear that the associations of alleles encountered in one generation depend on the manner in which individuals of the preceding generation associated for the purpose of procreation. For alleles with identical frequencies, the genotypic associations can be very different. Thus the frequency of the allele R of the Rhesus system would be one-half in a population comprised entirely of Rr heterozygotes (which would mean that the preceding generation had been composed of two types of individual only, RR and rr, the first of which had systematically procreated with the second). However, the frequency of the R allele would also be one-half in a population composed of one-half RR homozygotes and another of rr homozygotes (as happens if the RR procreate among themselves, and the rr among themselves).

But such extreme cases are rarely observed. It is noteworthy that, in practice, as soon as the allele frequencies are known, the genotypic frequencies can be predicted as follows with reasonable accuracy:

— *the frequency of the homozygotes (aa) is equal to the square of the frequency of the gene a,*
— *the frequency of the heterozygotes (ab) is equal to twice the product of the frequencies of gene (= allele) a and gene b.*[1]

1. Readers who are undeterred by simple probabilistic reasoning may wish to see how these results were achieved:

Thus, in a population where a single function is regulated by two genes a and b, the first with a frequency of 1/10, the second with a frequency of 9/10, the three possible genotypes have the following frequencies: 1/100 for aa, 18/100 for ab, 81/100 for bb.[2]

This result was established in 1908, a few years after the rediscovery of Mendel's laws, by Hardy, an English mathematician and Weinberg, a German biologist.

This rule is the culmination of a mathematical argument requiring numerous hypotheses, which together constitute a "model" of population structure, designated by the term panmictic; such panmictic populations may seem far from realistic. One must, in particular, hypothesize that marriages occur at random: that is, that the genes under study do not influence choice of mate. In human populations where this choice is usually subject to numerous rules, one might think that the model bears no relationship to reality and that the Hardy-Weinberg rule is inapplicable. In fact, each time that a verification could be carried out, it was found that the observed genotypic distribution agreed extremely well with that suggested by the model: the deviations from the theoretical expectation are usually no greater than the statistical sampling error. This

---

For a child to be homozygous (aa), two events are necessary, that an a gene be transmitted by both his father and his mother. The probability of each of these events taken individually is equal to the frequency of the a gene. The probability that the two of them would occur simultaneously is equal to the product of their probabilities. The probability of receiving two a genes is therefore equal to the square of the frequency of a (for example, 4 percent, if this frequency is 20 percent.)

For a child to be heterozygous (ab), it is necessary:

— either that his father have transmitted an a gene and his mother a b gene, an event with a probability which is equal to the product of the frequencies of the a and b genes,

— or that his father have transmitted a b gene and his mother an a gene, an event with the same probability as the preceding one.

The probability that one of these incompatible events should occur is equal to the sum of their probabilities. The frequency of the ab genotype is therefore equal to twice the product of the frequencies of a and b (for example, 12 percent if these frequencies are 20 percent and 30 percent).

2. Note that the two islands A and B which we imagined in chapter 1 did not conform to this model. They were not in "equilibrium." In island A, for example, the frequency of the d gene is 20 percent so that the 3 genotypes should have been distributed as follows: 4dd, 32Dd and 64DD and not, as we had assumed, 0, 40, and 60. A difference of this kind can very easily arise due to migration, but it cannot persist for several generations.

means that the choice of partner, no matter how deliberate, has no noticeable collective effect on the distribution of genes in the next generation, with the exception of rare particular cases.

In spite of its somewhat abstract formulation, this law is not just a mathematical curiosity for biologists who like to dabble in mathematical theory; it allows the correct interpretation of some seemingly paradoxical facts.

Let us return to the example of a population where the frequencies of the alleles R and r of the Rhesus system are both equal to 1/2. According to the Hardy-Weinberg rule, the frequency of (rr) homozygotes, that is of individuals with the Rhesus negative phenotype will be $(1/2)^2 = 1/4$; the frequencies of RR and Rr individuals with a Rhesus positive phenotype will be $(1/2)^2 = 1/4$ and $2(1/2)(1/2) = 1/2$ respectively, that is a total of 3/4. Although the two alleles have equal frequencies, the "positive" individuals are three times as numerous as the "negative" ones. Here again, we notice a divergence between observations made at the phenotypic level and those made at the genotypic level.

"Healthy" and "Defective" Families

The Hardy-Weinberg rule is valid not only for the blood group systems; it also applies to those diseases whose genetic mechanism has been established, in particular to the many "inborn errors of metabolism." One of the most frequent of these, in Europe and North America, is cystic fibrosis which affects, on average, one child in 2,500 (this proportion is as high as one child in 400 in certain regions) (Bois and Feingold et al.). This disease is characterized by various syndromes all of which have in common an abnormally high concentration of sodium chloride in the sweat. This disease is very debilitating and, so far, its treatment has been less than effective; the prognosis, in medical terms, is "reserved." The study of its transmission within families has shown that it is due to the presence in double dose, or in homozygous state, of a certain allele that we will designate as $m;$ individuals who have received only one copy of this allele, the Nm heterozygotes, where N designates the normal allele, are completely unaffected; it is even impossible to detect the presence of the deleterious allele in their genotype because it is completely camouflaged by the presence of the normal allele. The proportion of children affected, 1/2,500, represents the frequency of homozygotes *(mm)*; applying the Hardy-Weinberg rule, we conclude that the frequency of the *m* allele is equal to the square root of this proportion, that is 1/50 or 2 percent; the frequency of the normal allele *N* is therefore 98 percent.

Let us now do a calculation that will allow us to reach a very disturbing result which goes counter to many accepted notions. The frequency of the heterozygotes *(mN)*, the unaffected carriers of the deleterious allele, can also be calculated from the Hardy-Weinberg rule. We get $(2)(2/100)(98/100) = 3.9$ percent: thus, almost 4 percent of children; i.e., 1 in 25, are "carriers." The latter are therefore 100 times more numerous than the affected children.

People's usual attitudes toward supposedly "defective" families should therefore change drastically. In a country such as France, with a population of more than 50 million, two million are carriers of the *m* allele, while the number of people afflicted with the disease, if they were to survive, would be less than twenty thousand. The vast majority of the alleles for this "defect" (98 percent in fact) are part of the heritage of families which appear healthy, but only because chance has hitherto spared them the birth of a homozygous child.

Even if no case of cystic fibrosis has been recorded in our family, each of us has one chance in 25 of being the carrier of an *m* allele. Although it seems paradoxical, this very high risk is perfectly compatible with the low frequency of births of affected children; in fact, the probability that I am a carrier is 1/25, the probability that any one of my sperm is a carrier of the *m* allele is therefore 1/50, since only half of them receive it while the other half receive the normal allele *N*. The same probabilities apply to the partner with whom I procreate. The probability of a child developing from an egg and a sperm that are both carriers of the *m* allele is therefore indeed $1/50 \times 1/50 = 1/2500$.

The rarer the defect under study, the more striking this paradox becomes. Take, for example, the case of a condition that modern medicine can cure: phenylketonuria. This is caused by a gene designated traditionally by the letter p; this gene is recessive; in other words, only those children with two copies of the gene (pp homozygotes) are affected. Their metabolism is incapable of performing one of the innumerable chemical reactions necessary to its proper functioning. A biochemical substance called phenylalanine, which would normally be consumed or metabolized (and which is consumed when the normal gene is present), accumulates in the tissues and fluids of the body, including the brain, causing brain deterioration, severe mental retardation, and eventually death. This process is now fully understood; an effective method for countering the effect of this defect has been developed: it consists of limiting the amount of phenylalanine in the child's body to the required minimum by means of an appropriate diet.

The frequency of births of affected children is significantly lower than in the case of cystic fibrosis: about 1 per 11,000 in France (i.e., about 50 cases every year). The frequency of the gene p is therefore equal to the square root of 1/11,000, which is 1/105 or 0.95 percent. The Hardy-Weinberg rule shows us that the frequency of healthy carriers of this gene, the heterozygotes (Np), is 1.9 percent, that is, one in every 52 individuals, or 210 times greater than the frequency of affected children. The number of carriers in France is therefore approximately one million. Therefore, this seemingly rare defect, which is represented by only 50 or so new cases each year, is in fact present in a very large number of families.

Here, the abstract mathematical terminology is more than just a game for the initiated; it facilitates the investigation of a practical problem that is inaccessible to straightforward observation. For instance, if we return to the duality on which we insisted in the preceding chapter: only phenotypes can be observed directly while the underlying reality, on which future generations depend, concerns genotypes. To go from one to the other, a theoretical model is necessary. The use of mathematical terminology provides the only means of manipulating these models. It is, of course, necessary in every instance to check that this exercise does not lead to the study of an imaginary world. In the case of the Hardy-Weinberg rule, this is easy to do. It is less easy when we are dealing with more complex models which attempt to incorporate the numerous parameters that characterize the real world. An example of such a complex model is that describing genetic drift, which plays a major part in certain current attempts to explain evolution.

Random Changes in Gene Frequencies: Genetic Drift and Founder Effects

We saw that the genetic heritage of a child is the result of two lotteries, one of which governs the choice of half the genetic complement of the father, the other half the genetic complement of the mother. If we consider not a family, but two successive generations within a population, we notice that the genetic heritage of generation 2 is constituted by genes drawn at random from the heritage of generation 1.

Consider the genes determining a particular elementary function, for instance a blood group; each individual has a pair of such genes. Assuming that the size (N individuals) of successive generations remains constant, the $2N$ genes of the second generation will be copies drawn at random from the $2N$ genes of the first generation. Because of the random nature of this process, a particular gene with a frequencq of p_1 in generation one is likely to have a different frequency, p_2, in the next generation. A simple probabilistic argument, which coincides with intuition, shows that the difference between p_1 and p_2, that is the variation in frequency between two successive generations, is likely to increase as the value of N decreases. The famous "law of large numbers" indicates that if, on the contrary, N is very large, this difference in gene frequency between generations is close to zero.

Imagine two couples on a deserted island and assume that the two men are homozygous RR for the Rhesus blood group system, and the two women heterozygous Rr. The frequency of the R gene is therefore $p = 3/4$. Each couple has two children; five different genotypic distributions are possible:

— all 4 children are *RR*, hence $p = 1$
— 3 children are *RR*, 1 is *Rr*, hence $p = 7/8$
— 2 children are *RR*, 2 are *Rr*, hence $p = 3/4$
— 1 child is *RR*, 3 are *Rr*, hence $p = 5/8$
— all children are *Rr*, hence $p = 1/2$

It is easy to show that the probabilities of each of these five distributions are 1/16, 4/16, 6/16, 4/16 and 1/16 respectively. The frequency of *R* remains constant in only 6 cases out of 16; and in 1 case out of 16, it reaches a frequency of one, which means that the *p* gene is eliminated and that this elimination is permanent, at least for as long as the population remains isolated.

A similar dispersion of frequencies will occur in the next generation, therefore all possible genotypic compositions in the third generation must be calculated for *each* of the possible second-generation outcomes listed above. The process ends only when $p = 1$ and the *R* gene is fixed, or when $p = 0$ and the *R* gene is eliminated.

If, instead of two couples, we had envisaged a population of 50 couples, all the men being *RR* and all the women being *Rr*, the number of possible cases for the generation of children produced by them would have been 101, the possible values of the frequency *p* would have been 1, 199/200, 198/200, . . . , 1/2. This time however, the extreme values 1 and 1/2 would have had only the infinitesimally small probability of $(1/2)^{100}$, that is one chance in 10^{30}, which amounts to a practical impossibility. On the other hand, the values close to the initial frequency of 3/4 would have had a much higher probability: in 9 cases out of 10 the frequency of the *R* gene in the children would have been between 0.80 and 0.70.

This gradual change in genetic structure, occurring by chance, without explicit cause, is described as the "drift" of the population.

With each generation, the phenomenon recurs without any tendency for the frequency either to return to its initial value or to approach some limit value. In the long run, this erratic drift can produce only two possible results: either the frequency becomes 0, in which case the gene has disappeared, or it reaches unity which means that all the other genes for this function have disappeared and that the population has become homogeneous.

Gradually, as the composition of the gene pool is transformed by chance, the population evolves. However, this process is extremely slow; it is possible to show that the achievement of a significant change requires the passage of as many generations as there are individuals in a population. In a group of 100 people, the effects of genetic drift would not be felt for a few thousand years. Even then, the population would need to have been totally isolated, with no new genes being introduced through migration to interfere with the process. It is partly because of considerations such as this that geneticists are especially interested in "isolates," small human groups who have lived in almost perfect genetic isolation for a long period of time, usually for cultural rather than for geographic reasons.

Of course, the size of any group never remains constant and population sizes sometimes vary considerably. In practice, significant genetic drift occurs only during periods of reduced group size, for instance, due to some catastrophe such as an epidemic or to fragmentation of the population into several subgroups that remain distinct from that point onward.

This kind of population subdivision has occurred frequently in the course of the history of human populations. Its impact on the genetic heritage is denoted by the term "founder effect." Most new human populations were founded in this way, by a small number of individuals who had left the ancestral group into which they were born, either in search of a better life or in revolt. Their genes, which become the biological heritage of the new population, are no more than a sample of the original group's genes; the fewer the founders, the less representative this sample will be.

We were thus able to compare two Tuareg groups from the Southern Sahara (Chaventré and Jacquard), the Kel Dinnick, descendants of the noble Tademaket tribes, who reigned over the entire region of Adnar des Iforas until the seventeenth century, and the Kel Kummer, whose ancestors, members of those same tribes, seceded at that time and gradually came to dominate the area. The

small number of warriors who revolted and founded the new Kel
Kummer group represented only a very limited sample of the genetic heritage of their ancestors. André Chaventré (1974) showed,
after extensive genealogical study, that 40 percent of the genes now
carried by the Kel Kummer came from a mere 5 founders and 80
percent from 15 founders. Blood samples made it possible to establish, for many different blood group systems, the genetic structure of the Kel Dinnick and their relatives the Kel Kummer. In
some cases, there are considerable differences. For example, for
the immunological system "HL-A," the genes which are in the
majority in one group are absent or rare in the other and vice versa.
Chance alone is responsible for these differences and it would be
futile to look for their causes.

Migration: An Essential Factor in the Transformation of Populations

All our calculations of genetic drift were based on the following
assumption: the complete isolation of the group over the period
covered by the generations under study. When this isolation is absolute, the long-term effect of the drift is to homogenize the population: for each elementary function, only one gene survives. Ultimately, all individuals in the population become homozygous: they
are genetically identical, as are monozygotic twins.

In practice, this point is reached only in experimental laboratory
populations. The experiments in question are carried out for the
sole purpose of generating homogeneous groups of animals. For
instance, in mouse breeding, homozygous strains are obtained by
cross-mating brothers and sisters over many generations.

In human populations, a variety of circumstances are likely to
occur which break the genetic isolation at some point during the
long period necessary for the drift to significantly increase the
homogeneity of the group. Moreover, even a tiny trickle of immigration is enough to neutralize the effects of this drift.

Consider, for example, a population composed of 50 individuals
in each generation. It is possible to demonstrate that, to homoge-

nize half of the elementary traits that initially showed some diversity, 70 generations or more than 1,500 years are required. It is not probable that absolute isolation could be maintained for so long a period of time. If we assume that even one immigrant enters the population at each generation, the proportion of traits to become homogeneous drops from 50 percent to 10 percent. Each immigrant contributes "fresh" genes which spread throughout the group and replace those that have been eliminated by drift. If even a very slight current of immigration is maintained, the tendency toward the erosion of genetic variation by drift will not lead to the homogenization of the population.

This genetic impact of immigration, which is far greater than expected, given that such a small number of migrants is involved, is further increased by the fact that migrants very often have more children than average. Whether for biological reasons (migrants are the result of a certain selection from within their original populations) or psychic ones (they try to recreate a familiar world around them), this higher fecundity is often observed. An extreme example of this was found in a small tribe of Jicaque Indians in the Honduras, studied by ethnologist Anne Chapman (1971). The genealogies that she established trace the genetic history of the group since its foundation a century ago by seven people (4 men and 3 women); immigrants constituted only about 5 percent of the total number of married people in each generation. However, 29 percent of the genetic heritage of children born in recent years is composed of genes from those immigrants while only 71 percent of it comes from the historic "founders." The higher fecundity of the immigrants gave them a far greater genetic role than their actual number suggested.

Often, members of a group considerably underestimate the intensity of the influx of fresh genes: this was found to be true of a Protestant community in Normandy, which has been isolated since the Reformation in a predominantly Catholic region. Until the Vatican II Council in the 1960s, mixed marriages were extremely rare, being the object of intense general disapproval. The tightness of the marriage market for Protestants meant that some marriages between relatives were inevitable. A feeling of strong consanguinity is widespread in this group, especially since many families have the same family name. The complete genealogy of this community is now at our disposal, thanks to research carried out by Martine Segalen (1973). She found that more than half of the 217 people present in this community in 1950 were descended from the same founder couple, a third were descended from another founder, a finding which may seem to justify their feelings of extreme isolation, strong consanguinity, and, therefore, of genetic impoverishment. In fact, in each generation, a few spouses came from outside the community. It was possible to calculate that, on an overall basis, the proportion of "new" genes being added to the collective heritage in less than 25 years was greater than 20 percent. Therefore, this group, contrary to its belief, was not in a state of total genetic isolation and in danger of becoming gradually impoverished, but was in fact benefiting from a constant input of genes from outside, which have transformed its biological makeup.

Loading the Dice in the Mendelian Lottery: Natural Selection

We have shown that the way in which a generation of children is derived from a generation of parents can be compared to a series of lotteries: each child, for each elementary trait, receives two genes drawn at random from the genetic "pool" of both parents. But this chance process does not necessarily give an equal chance to all the parental genes. If a particular gene causes a decrease in fertility or a lowering of resistance to disease, the individual who is carrying it will participate less often than others in the various lotteries and therefore his biological heritage will be less well represented in the next generation. The relationship between genetic endowment and the capacity to survive and procreate is the basis for the phenomenon of *natural selection*.

This word "selection," because it is so widely used, is likely to cause confusion. It is necessary to indicate precisely in what sense we are using it. First of all, it denotes the deliberate attempts made by breeders to modify particular traits in plant and animal species.

Observation has shown that certain crosses have well-defined effects on the progeny; a systematic technique has been developed, which makes it possible to have a strain evolve in a desired direction; this merely requires skillful selection of the individuals to be used for reproduction; it involves artifice, some degree of manipulation of the natural order of things: by definition, selection is artificial. On the contrary, when we speak of the effects of *natural* selection on various genes, we use the term "natural selection" as it was introduced by Darwin. He chose it to underline the fact that the mechanism responsible for the spontaneous transformation of species relies on the same material as that used by breeders: the diversity among individuals for the trait in question. However, in the natural process we are now studying, this diversity is no longer simply the raw material used by an external agent, the breeder; rather, by itself, it influences the reproductive power of individuals and causes it to vary, thus bringing about a change in the frequencies of various genes from one generation to the next. In other words, it causes a natural evolution of the biological heritage.

Let us return to our example of a population where the frequency of the R gene for the Rhesus system is $p = 3/4$. When the conditions specified by the Hardy-Weinberg rule are fulfilled, 1/16 of individuals have the homozygous genotype (rr) and manifest the negative trait. It is known that while a Rhesus negative woman is pregnant with a Rhesus positive child, her body synthesizes antibodies which could, in a subsequent pregnancy, agglutinate the red blood cells of a Rhesus positive fetus, and could therefore, in the course of delivery of the baby, be the cause of hemolytic disease in the newborn, a disease that, in the past, was often fatal. In our example, this risk is incurred in three out of every four pregnancies of Rhesus negative women, i.e., women of the rr genotype. Therefore, these women have, on the whole, fewer children than average, thereby lowering the frequency of the r gene. In a population where the risk of death, due to this kind of mother-child incompatibility, is 20 percent, it can be calculated that the frequency of the r gene would diminish by 0.2 percent at each generation. Therefore, in spite of the high risk, the transformation of the genic structure is very slow. According to this hypothesis, more than six centuries would be required to lower the frequency of the *r* gene from 25 percent to 20 percent.

In the case of a recessive "defect" such as cystic fibrosis or phenylketonuria, the death of affected children is likewise very slow in bringing about a reduction in the frequency of the gene responsible for the particular trait. Assume that, in the past, all those individuals afflicted with phenylketonuria, the *pp* homozygotes, died before reaching reproductive age but that the reproductive potential of carriers, i.e., *Np* heterozygotes, was in no way reduced. The frequency of the *p* gene, which is currently estimated at 0.95 percent, would have been reduced to 0.90 percent, a very slight improvement, only after 6 generations, or a century and a half, and to 0.50 percent after 95 generations, or more than twenty centuries.

It may seem paradoxical that a gene which causes death should disappear so slowly. In fact, this could have been predicted from the Hardy-Weinberg rule: the *p* genes which are expressed in homozygous individuals are indeed eliminated, but they represent only a tiny minority. For example, in the entire population of France, natural selection can affect only about 10,000 *p* genes, while almost a million of them escape it, sheltered by the effective camouflage provided for them by the normal gene with which they are associated in heterozygotes.

Without going into excessively complex technical details, we will now finish by dwelling briefly upon one frequently misunderstood aspect of natural selection: it acts on individuals and not on genes. It is a mechanism which operates at the phenotypic level. An individual's ability to resist the various aggressions of his environment, to survive and to procreate is a function of the many characteristics of his phenotype. His overall success or failure will affect all the genes of which he is a carrier; of these, some are beneficial and others are harmful; their fate in passing from one generation to the next will be less a function of their own characteristics than of the global characteristics of the individuals who carry them. A particular gene with a positive effect may nevertheless disappear if it is associated with a lethal gene; another, which lowers resistance to disease, will be increased if it is associated with genes that cause higher fertility. For the sake of simplicity, we have no choice but to consider each trait in isolation; however, it must be borne in mind that this is a simplistic model, and very far from the complexity of reality. We will return to this difficulty when we deal with the diverse theories that attempt to explain the evolution of species: Darwinian, neo-Darwinian and non-Darwinian theories.

The Prospect for Planned Changes in the Genetic Heritage

The collective genetic heritage of a group constitutes its biological wealth, its only truly lasting possession. This heritage, in being transmitted from one generation to the next, is transformed spontaneously due to mutations, migration, the element of chance introduced by the multiple Mendelian lotteries ("drift"), and differences between individuals' reproductive capacity ("selection"). Could we not hope to modify this heritage at will, now that we have begun to understand it?

The aim of "eugenics" is to define strategies or techniques allowing us to do our best to ensure that these conscious changes are beneficial. This very aim makes it necessary to think in terms of genotypes: it follows that all discussions in this regard should take into account the concepts and results of "population genetics." It is noteworthy that these two areas of research, "eugenics" on the one hand and "population genetics" on the other, have developed separately, and that the former has failed to take account of the contributions of the latter. Most of the statements made by eugenicists are in total contradiction with the elementary principles established by population genetics. An obvious illustration of such contradiction is seen in the misinterpretations made, not of the difficult concept of species improvement, but of a more easily defined concept, that of the danger of species deterioration. The next chapter deals with the major fears expressed in this regard.

THREE

Future Prospects for Our Genetic Heritage: Real and Imaginary Dangers

MAN LIVES in a world of his own making. Without realizing it, we have transformed, among other things, the conditions under which genes are transmitted from one generation to the next. Some people fear that by pursuing specific objectives, such as curing sick children, energy production, or social stability, we may be interfering with the natural order and triggering a process that, in the long run, could lead to catastrophe. Fears regarding genetic deterioration are often voiced; we will examine these fears from the perspective provided by population genetics to determine what exactly are the dangers created by such factors as the "dysgenic effect" of modern medicine, the consanguinity of procreating couples, and the mutational effect of irradiations or of certain chemical products.

An Unfounded Fear:
The "Dysgenic" Effect of Medicine

Doctors are often criticized in terms somewhat like the following: "In treating a child with a genetic defect, and by curing him, you enable him to lead a normal life; you are totally successful if you make it possible for him to have children. But this very success leads to a terrible danger: the harmful genes that this child received from his parents are now going to be transmitted on instead of being eliminated as nature intended. Gradually, the collective genetic heritage is going to be encumbered by these harmful genes; your intervention, undoubtedly beneficial in the short term, will actually lead to catastrophe in the long run."

This argument seems so clear that it convinces many doctors and causes them to voice concern about the long-term consequences of their treatment. The example most frequently cited to demonstrate the boomerang effect of some medical achievements is diabetes. For the past fifty years, insulin treatment has been successful in allowing diabetics to lead a normal life and, in particular, to have children; during the same period, the frequency of this disease has increased: in countries with high medical standards, almost 4 percent of elderly people are affected and require treatment in order to survive.

Similarly, over the past twenty years, effective techniques have been developed for keeping hemophiliac children alive and allowing them to reach reproductive age. In countries such as Denmark, where precise medical statistics are available, the frequency of this disease had been remarkably stable for a long time. It now seems that this proportion is on the increase for the first time in some years.

It is arguments and facts such as these that have kindled fears about the "dysgenic effects" of medical progress or, to put it colloquially, about the "genetic contamination" of the human species.

Genetic Change: An Extremely Slow Process

The most notable characteristic of changes to the genetic heritage, and one which seems to escape the prophets of doom, is their extreme slowness: genetic change in an individual cannot become manifest until he produces offspring; the time unit in this domain is not the year but the generation, i.e., a quarter of a century roughly. It is in terms of generations that the time required for such a change to become manifest must be calculated.

Consider, for instance, a very serious genetic disease for which modern medicine has found a cure: phenylketonuria. We saw that this "defect," which, in France, affects one child in 11,000, is due to a recessive gene p. For about twenty years, or less than a generation, the effects of this p gene in individuals with a homozygous pp genotype have been eliminated. Special dietary treatment allows them to escape their natural fate, which used to be one of progressive brain degeneration and death before reaching reproductive age. Henceforth, they can lead a normal life and even have children. To be sure, they transmit a p gene to each of the latter; but, since this gene is rare in the population, their partner usually transmits a normal N gene and the children are unaffected. The probability of having an affected child is exactly 0.95 percent. The risk run by these families is therefore very slight, but is there not a risk of the collective genetic heritage becoming progressively overloaded with this p gene? Since a natural balance has been interfered with, has a long-term catastrophe not been triggered? To answer this question, we need to use the arguments about global genetic balances outlined in the preceding chapter.

The frequency 1/11,000 of homozygotes corresponds to the frequency 1/105 of the p gene. The fact that this gene has not disappeared long ago, in spite of the death of all children who had a double copy of it, suggests that mutations introduce new p genes at each generation, or, as is more likely, that heterozygotes have, or used to have in the past, some kind of advantage. Regardless of which hypothesis one accepts, because of the cure of affected children, there is now a different balance, and the frequency of the p gene will increase gradually; a method for calculating the number of generations required to double this frequency is presented in the appendix: this number is somewhat greater than 50. The doubling of the gene's frequency (leading to a quadrupling of the frequency of affected children) will therefore not happen for about 1,500 years. Therefore, because of medical progress, one child in 2,800, instead of one in 11,000 as at present, will need treatment.

However, this frequency, which is still quite low, will not be reached till the year 3500: does it make sense to worry about this danger when, in the meantime, humanity will have to confront problems so great as to jeopardize its very existence.

Of course, this slowness is partly due to the fact that we have chosen a relatively rare disease as an example. The more frequent the genetic disease, the sooner its treatment and cure will affect the genetic structure of the population; but, in all cases, the rate of change nevertheless remains slow. To demonstrate this, consider the case of the most widespread genetic disease in Europe: cystic fibrosis. It is possible to calculate that a complete cure, which unfortunately is far from being available at present, would lead to the doubling of the frequency of the c gene responsible for the disease in about 700 years. Around the year 2700, the number of affected children would be about 1 per 600 instead of the present 1 per 2,500. It is clear that this danger is far from imminent.

The process is more rapid with diseases such as hemophilia, whose genetic determinism is different from that of the two we have just examined, in that it is governed by the "sex-linked genes." These genes are located on the X chromosome of which males carry only one copy; while women have a pair of XXs, men have a dissymmetric pair, an X and a small chromosome Y, responsible for their masculinity. The gene for hemophilia, let us call it h, is recessive (like the r gene for the Rhesus system) and situated on the X chromosome. Its frequency in Europe is roughly 1/10,000. To be a hemophiliac, a woman would have to be a homozygote (hh), an extremely rare occurrence since its frequency, according to the Hardy-Weinberg rule, is $1/10,000^2$, or 1 per 100 million; but in the case of a man, one copy of this h gene on his single X chromosome is all that is required for the disease to manifest itself. The frequency of births of hemophiliac boys is therefore 1/10,000.

A simple calculation (see the appendix) shows that the curing of all those affected would lead to an increase in the frequency of the gene equal to 1/30,000 in each generation: in one century, the frequency of the disease would be doubled; but it would still affect only 1 boy in 5,000 and it would be another 1,000 years before this incidence would be greater than 1 boy in 1,000. Therefore, the negative consequences of medical progress become evident, in this case within a shorter time, but they remain limited; it can hardly be argued that there is an immediate problem.

This attitude may seem irresponsible and reprehensible: even if the genetic contamination is not to occur for a few thousand years, the risk may seem very real; we are responsible for the long-term destiny of our species; we do not have the right to let problems accumulate, even if their effects will not be felt till long after our own lifetime.

It is therefore best to evaluate this risk independently of the time factor. To do that, it is necessary to establish precisely what it consists of: we shall see that matters are less simple than they might seem.

The Significance of Genetic Changes

When the frequency of a gene is doubled, the frequency of individuals homozygous for this gene is quadrupled, since the latter is equal to the square of the former; simultaneously, the frequency of heterozygotes, carriers of one copy of the gene, is approximately doubled.

Now, as we have seen, one of the hypotheses proposed to explain the maintenance of a harmful gene is that it confers some advantage on heterozygotes. This is not just an academic hypothesis with no grounding in reality; in the human species, it has been shown to be probably true in at least one well-known case, that of sickle cell anemia, so-called because it is characterized by the "sickle" shape of the red blood cells. This disease is very widespread in those parts of Africa that are plagued with malaria; it has been found to be due to a gene S, which is responsible for the malformation of the red blood cells. Children who have two S genes, the SS homozygotes, almost all die of anemia; but heterozygous individuals who have received the S gene from only one parent, the other supplying a normal gene, seem to some extent protected against malaria (no doubt because the malformation of a portion of their red blood cells interferes with the normal developmental cycle of the parasite, *Plasmodium falciparum,* which is responsible for malaria). In this one case at least, the advantage associated with the heterozygous state seems a concrete fact.

It seems likely that this mechanism may also have applied, or may still apply, in the case of cystic fibrosis. The frequency of this disease is too high for mutations to have been able to compensate for the loss of c genes due to death of homozygous cc children. Such mutations do sometimes occur, but their frequency is always extremely low and is certainly much too low to explain the maintenance of the frequency of c at the level of 2 percent.

This frequency is very probably due to some advantage to heterozygotes (in the form, for instance, of better resistance to some diseases). Even a slight advantage, so slight as to be impossible to demonstrate clearly, would explain this frequency. For the present, this hypothesis can be neither proved nor refuted.

In every case where heterozygotes are thus advantaged, the detrimental effect on the population of affected homozygotes finds immediate compensation because of the increased fitness of heterozygotes.

Consider again the possibility that medical progress provides an immediate cure for all children afflicted with cystic fibrosis: these children, if they survived, would total about 20,000 over the whole of France. We have seen that, because of this very progress, seven centuries from now, the frequency of the c gene would have doubled, and that of homozygotes quadrupled: the treatment essential to their survival would therefore have to be given to 80,000 people (assuming a constant population). But simultaneously the number of heterozygotes would have doubled, the number of these "advantaged" individuals would have swelled from 2 to 4 million. It is impossible to predict whether the overall consequences of this change in the genetic balance would be positive or negative.

Note, above all, that the type of medical progress envisaged here would mean that cystic fibrosis would no longer be considered a "defect"; it would become a disease requiring treatment but, hypothetically, curable. The increase, from 20,000 to 80,000, in the number of people affected by it, would constitute not a genetic burden, but an economic one. Would it not be derisively small, compared to other economic burdens that stem from the imperfections of our social structures?

This process is no different from that which has been taking place since the dawn of humanity when *Homo sapiens* came to be; ever since, we have been inventing appropriate behaviors in response to the aggressions of our environment; we have not been content to wait passively for a genetic mutation. The invention of fire and the use of animal skins certainly prevented the elimination of children with genetic endowments that provided less efficient resistance to cold. This has, in the long run, transformed humanity's genetic heritage: as a result, we are no doubt more frail, in some respects, than we once were. However, to describe this frailty as a genetic deterioration is to go too far. It is in the very nature of our species to alter its natural environment to its own advantage: as soon as it was within our power, we ceased to passively endure the selection imposed by our milieu; to its aggressions and constraints we gave a cultural response, and not, as with other species, a genetic response; medical progress is nothing other than a continuation of this response; the invention of an antibiotic is no more "dysgenic" than the invention of fire.

Gene-Environment Interactions

Scientists wonder whether these "bad" genes, which medical progress now allows us to tolerate and to preserve for the future by neutralizing their negative effects, have always been harmful. One explanation for their presence in our biological heritage is that, formerly, even in the homozygous state, they conferred some advantage.

This seems to be the case with some forms of diabetes. This disease manifests itself in many different ways (going from almost complete absence of insulin synthesis to a slight increase in blood-sugar level); but, above all, it appears at widely different ages: while it affects less than 1 per 1,000 adolescents, it has been found to affect more than 4 percent of elderly people. Its genetic determinism is still the subject of some debate; it seems to involve the interaction of numerous genes which determine not the disease itself but a person's predisposition to manifesting it; the disease appears only when a person's intake of food exceeds a certain threshold of "richness," this threshold being itself determined by the genes that are present. One individual whose genes specify a strong predisposition to diabetes will not manifest the disease as long as his diet is sufficiently limited; another, though much less predisposed genetically, will be affected if his diet is excessively rich.

The actual frequency observed in any particular country, therefore, depends less on the genetic structure of the population than on its eating habits and food availability. The increase in the frequency of diabetes that has been noted in some industrialized countries over the past half-century is, therefore, not a consequence of medical progress allowing diabetics to leave more offspring (we saw that the time span is too short for such an effect to have become evident), but of changes in the diet, which has become increasingly abundant and rich. The higher incidence of diabetes is not, therefore, a signal of some change in the genetic heritage; it is merely the manifestation of some of its existing properties, which could not previously be observed.

The genetic combinations, which, under certain dietary conditions cause diabetes, seem, in our present circumstances, to be harmful. However, it is quite possible that, in time of famine or of food scarcity, these same combinations would have beneficial effects. Perhaps they favor a type of metabolism that allows the body to benefit more from those substances that are available. This possible reversal of the effect of the genes in question was suggested especially for Professor Neel (1962) of the University of Michigan at Ann Arbor. In preventing the disappearance of these diabetes-associated genes, medical progress does not, therefore, have a dysgenic effect. On the contrary, it is preserving a genetic capital which is doubtless useless or maladapted at present, but which may prove to be invaluable when we are less glutted with food.

What To Conclude?

Again and again, we will have occasion to reiterate this frequent conclusion in biology; the notion of good and bad in biological systems is a gross oversimplification and is not very useful when dealing with the complexity of living things; it is very rare that a particular genetic configuration can be deemed absolutely bad. Usually, the "context" determines the kind of judgment to be made, "context" here meaning both the genetic heritage as a whole and the environment: the S gene of sickle cell anemia is "bad" for homozygotes, it leads to their death; this same gene is "good" for heterozygotes, it protects them from malaria; the genetic combinations responsible for diabetes are "bad" for a person who is too well nourished, but they may be "good" for that same person in time of famine.

Given these facts, how can anyone claim that medical intervention leads inevitably to degeneration?

One may well wonder about the large impact of statements to this effect, even when such statements are made lightly by researchers who have not thought the issue through. This impact, and associated media coverage, is no doubt partly due to the widespread taste for all that is catastrophic and apocalyptic. Also, it is doubtless true that, by attacking medicine in the name of genetics, one can easily give oneself the impression of going beyond the immediate goals of medicine and of seeing further into the future. Above all, however, this attitude reveals the intrinsic attraction of any concerted policy of "improvement" of the race or species; the fear of a dysgenic effect of medicine is merely the negative side of the hope for a positive eugenic program.

An Ill-Defined Danger: Consanguinity

In most cultures, marriages between close relatives are prohibited. Unions considered to be incestuous are the object of taboos that are usually very strict. This attitude is commonly attributed to a fact that could have been observed in any society: children born of such unions have less chance of surviving and they have more defects. In prohibiting them, populations are believed to have been trying to prevent genetic deterioration and to have been practicing a kind of "eugenics." Is this really the origin of this behavior?

The Definition of Relatedness

Two people are genetically related if they have one or more common ancestors or if one is an ancestor of the other. Though this definition seems clear it leads immediately to this paradox: each of us has two parents, four grandparents, etc. By continuing this duplication back about thirty generations to those of our ancestors who lived under Philippe le Bel, we reach a number close to one billion, which is much greater than the entire population of the earth at that time. The error that we made was in multiplying by two at each generation when, of necessity, some of our ancestors could be reached by several paths, paternal or maternal. In a small population, relationships gradually increase and become more intricate; after a few generations, everyone is related to everyone else. This overlap between genealogies is another aspect of the phenomenon of "drift" that we described in chapter 2. Needless to say, mathematicians have studied this question and, with the help of rather complex arguments, have succeeded in calculating the rate at which this process occurs, as a function of the number of people in the group, the rules governing marriages and the distribution of family sizes. Thus, Thierry Leviandier (1975) was able to prove that, in a population composed at each generation of 50 men and 50 women, with random mating and an average of two children per person, it would take only 11 generations for each individual to have 80 percent of the founders among his ancestors; with 500 men and 500 women, it would take only slightly longer: 15 generations.

This theoretical result is confirmed by studies of "isolates": in the Kel Kummer Tuareg tribe, whose genealogies back to the seventeenth century were reconstituted by André Chaventré (1974), it was found that any two individuals in the present population have at least 15 ancestors in common. Part of the genetic complement of these "15 founders" is to be found in all the present members of the tribe, without exception . Therefore, they are all related.

The idea of "relatedness" can only be useful if one specifies the number of generations in which one looks for common ancestors: it is therefore necessary to reach agreement on this point before

comparing different populations. Very often, only 3 or 4 generations are taken into account (particularly in some European countries, where studies are based on records of dispensations given by
the Roman Catholic Church); the search, in such cases, is limited
to the identification of common ancestors for second cousins or, in
some cases, for third cousins. This is the case, for instance, with
the classic work of J. Sutter (1968) on this subject. However, it is
important to remember that relatedness is not an intristic characteristic of two people, it is merely a characteristic of the information that could be gathered on their genealogies.

Quantifying Relatedness

The only biological consequence of relatedness between two individuals, A and B, is that they have part of their genetic heritages
in common: a certain number of genes that have come down to
them from their common ancestors. The simplest case is that of
two half-brothers with, for instance, a common father and two different mothers (figure 2). At each conception, the father, F, supplied half of A's genes, then of B's. However, the half transmitted
to A is, of course, not exactly the same as that transmitted to B
since the spermatozoa involved result from two different "lotteries." Some of the genes from F that were chosen to make the sperm
that produced A were, by chance, also chosen for the sperm that
produced B, others were not. Chance therefore plays a fundamental role; the measure of relatedness, to be faithful to the nature of

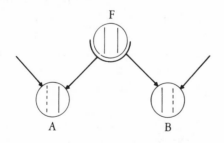

Figure 2

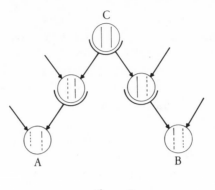

Figure 3

things, must take it into account. The best definition of such a measure was proposed by G. Malécot (1948). Consider any basic genetic trait; for this trait, both A and B possess a pair of genes; choose at random one gene from each of these pairs; the probability that these two genes are a copy of the same gene from one of their common ancestors is, by definition, their "coefficient of kinship."

In the case of two half-brothers, it is easy to calculate this probability. It can be shown that the two genes chosen at random have 1 chance in 8 of being a copy of the same ancestral gene; in short, their coefficient of kinship is ⅛.[1] When two genes are copies of the same ancestral gene, they are described by geneticists as "identical." The probability of two genes, one from A and one from B, being identical decreases the further the common ancestor is removed from them: two people with just one common grandparent (figure 3) will have a coefficient of kinship of 1/32; with a single common great-grandparent this coefficient becomes 1/128.

If there are several common ancestors, the contribution of each of them must of course be added: the coefficient of kinship for two brothers, with a father and a mother in common, is ¼, for two

1. Specifically, the gene chosen from A has one chance in two of having been supplied by F, likewise with the gene from B and there is one chance in two that F transmitted the same gene to both A and B; finally, there is one chance in eight that the two genes chosen at random are identical.

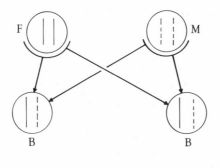

Figure 4

first cousins with two grandparents in common it is 1/16, for two second cousins it is 1/64, etc.

It is to be noted that the coefficient of kinship is the same, ⅛, for the uncle-niece relationship, the double first cousin relationship (where two brothers marry two sisters), and for the relationship between half-brother and half-sister. However, marriage in the last instance is considered incestuous in France, while in the other two cases the necessary authorizations are easily obtained.

With very complex relationships, the calculation of these coefficients can become extremely laborious; the use of a powerful computer is obligatory, so numerous are the possible paths that the genes of the various "founders" may have followed (several hundred thousand in the case of the genealogies of the Kel Kummer Tuareg) (Nadot and Vayssex).

Marriages Between Relatives and Abnormalities Among Their Offspring

When two relatives produce a child, they run the risk of transmitting to their child two genes that happen to be copies of the same gene from one of their common ancestors. These two "identical" genes are, of necessity, exactly the same (unless a mutation has occurred, but this is so rare an event that it can be ignored); the child is therefore homozygous for the corresponding trait.

Even if the father and mother are not related, the child may very well be homozygous; the Hardy-Weinberg rule tells us that the probability of this occurring is equal to the square of the gene frequency. When they are related, this homozygosity may result, not only from the chance meeting of two genes with the same function, but also from the transmission of two identical genes, an event with a probability equal to the "coefficient of kinship" as we have defined it.

The only biological consequence of marriage between relatives is this increase in the proportion of homozygous traits in the children. We saw that some diseases were due to recessive genes, that is, their ill effects become expressed only in the homozygous state; the likelihood that these diseases will become manifest is therefore greater in consanguineous families. The coefficient of kinship that we described will allow us to calculate this risk exactly.

According to our calculations in chapter 2, the frequency of the c gene for cystic fibrosis is in the region of 2 percent. For an unrelated couple, the risk of giving birth to an affected child is the square of this frequency, 4 in 10,000 or 1 in 2,500. However, a couple who are related, for instance first cousins, run an additional risk: both father and mother may, with a probability equal to their "coefficient of kinship" (1/16 in this example), transmit two "identical" genes to their children; the probability that the ancestral gene of which these genes are replicates was a c gene is equal to its frequency, 2/100; therefore, the probability that the child will be cc, due to the relatedness of his parents, is $1/16 \times 2/100 = 12/10,000$; the overall probability that first-cousin parents will produce a child with cystic fibrosis is about 16/10,000, or four times greater than if they were not related.

It is easy to see that this increase in risk becomes greater the rarer the disease being considered. In Europe, for example, the frequency of albino children is about 1 in 20,000. This disorder is due to a recessive gene, say a: the aa homozygotes are unable to synthesize sufficient melanin, the pigment required for coloring the hair and skin as well as the iris of the eye, which explains their characteristic "colorless" appearance. The frequency of the a gene is, according to the Hardy-Weinberg rule, 1 in 140. The same line of argument as that applied in the case of cystic fibrosis shows that, for a couple who are first cousins, the risk of having an albino child is as high as 1 in 2,000, ten times greater than for unrelated couples.

These calculations are true for all recessive diseases. On the contrary, in the case of diseases linked to a dominant gene, which are expressed both in heterozygotes and in homozygotes, in theory the effect is the opposite, but, since the genes in question are at a very low frequency, the decrease in the frequency of the disease due to relatedness is negligible.

In practice, the sequence of observation and argument is often the reverse of that we have just followed. Usually, the principal aim is not the prediction of offspring phenotypes for related couples. Rather, one attempts to infer a possible genetic basis for a disease because of its increased frequency among the offspring of related couples. Thus, in France, the proportion of marriages between first cousins is, on average, about 2 per 1,000; but hospital statistics show that it is as high as 8 per 1,000 in the group of families where a case of cystic fibrosis has occurred. A difference of this magnitude is a sign that this disease is very probably linked to a recessive gene; a detailed study of its mode of transmission within families then makes it possible to elucidate this mechanism (Frezal et al. and Robert).

Related Parents: Offspring Phenotypes

The various diseases labelled "genetic" constitute an exceptional category. Most human traits are in fact not produced by a simple mechanism depending on a single pair of genes; they result from numerous interactions involving both the environment and a multiplicity of genes. When parents are related, the degree of homozygosity for the genes involved is likely to be higher in their offspring and to affect the expression of the trait; however, in spite of extensive research aimed at defining this effect, few clear-cut results are available. Once again, it is almost impossible to infer the genotype from observation of the phenotype.

The most precise data available are contained in a study carried out, after the last war, in Hiroshima and Nagasaki, by the Atomic Bomb Casualty Commission (Schull and Neel 1965). This study showed a correlation between some offspring traits and relatedness of parents, but these correlations were extremely weak, close to what statisticians call the "threshold of significance."

For instance, in a total of almost 70,000 cases, N. Morton (1961) found that the average weight of children at birth was 3,046 grams when the parents were first cousins, 3,074 grams when the parents were unrelated; the length at 8 months was 68.73 centimeters in the first case, 68.96 cm in the second; the chest circumference at the same age was 42.68 cm and 42.77 cm, respectively. Consanguinity therefore causes some reduction in body measurements, but the differences are so small that they become apparent only when a huge sample is studied.

An identical result was reported when the same Japanese sample was tested for the effect of consanguinity on intellectual performance. We will return later to the problem of defining measures to characterize "intelligence" or the various components of intelligence. For the present, we simply note that the results of the various tests chosen by W. Schull to evaluate the intellectual aptitudes of Japanese children were slightly lower when the children were related; the difference between the children of first cousins and those of unrelated couples was, on average, 2.5 percent.

The differences observed are so slight that it is difficult to consider them as final proof of the harmful effects of consanguinity.[2] The feasibility of comparing a group of consanguineous families with a group of nonconsanguineous families is, in fact, not beyond dispute. Some of the differences could be due as much to the social or economic environment as to the genetic heritage. Moreover, what is true of one population may very well not be true of another, as studies on infant mortality have made clear.

---

2. And, above all, the methodological difficulties are so great (in particular, that of eliminating in the comparison the social and psychological factors associated with consanguinity).

Relatedness of Couples, Infant Mortality, and Sterility

One of the first serious studies of the effects of consanguinity on infant mortality was carried out by Jean Sutter and Léon Tabah (1951) in the French departments of Loir-et-Cher and Morbihan. Using the concept of "perinatal mortality" and relying on some very elaborate demographic techniques which make it possible to distinguish between deaths due to exogenous causes and those due to endogenous causes, especially genetic ones, they found that the risk of death was greater in the descendants of first cousins, by 90 percent in Loir-et-Cher, by 170 percent in Morbihan. These very high percentages have been widely used in support of theories about the ill-effects of consanguinity. However, the major study carried out in Japan by the Atomic Bomb Casualty Commission led, some years later, to rather different results: the increase in perinatal mortality among the offspring of first-cousin couples compared to that of unrelated couples was 27 percent in Hiroshima and 9 percent in Nagasaki. The lack of agreement between the two sets of results indicates that the results of such studies must be interpreted with the greatest caution.

The difficulty of insuring genuine comparability between samples of consanguineous and unrelated couples has already been mentioned. The smallest bias in the representation of the various social, ethnic, or professional categories may deprive the findings of all biological meaning. In an effort to eliminate these biases, J. Sutter and A. Georges carried out a particularly detailed study in two valleys in the Vosges, using as control couples the brothers and sisters of each of the first cousins whose children were examined (Georges and Jacquard 1968). In this case, the increase in mortality due to consanguinity was 23 percent.

A child may be adversely affected by greater genetic homogeneity long before birth; if an embryo has very serious defects, it may be eliminated very early in the pregnancy. Consanguinity must, if it is harmful, cause a higher frequency of spontaneous abortions and sterility.

Here again, the results of the various studies differ considerably, but everywhere substantial increases in the frequency of sterile couples were observed as a function of consanguinity: in Loiret-Cher in 1952, the rate of sterility was 6 percent in unrelated couples, while it was 10 percent in consanguineous couples (results for first and second cousins combined); in the Morbihan, these frequencies were 5.6 and 8.4 percent respectively, in the Vosges in 1968, 4.6 and 6.9 percent.

These results, in spite of their lack of precision, are in agreement with various findings suggesting that intrauterine death in early pregnancy is very frequent (no doubt about 50 percent of all conceptions) and plays a major role in the elimination of embryos afflicted with defective genetic endowments.

Summary

In the end, we are left with a group of observations, none of which is entirely decisive, but which are all consistent with the results predicted by theoretical models: a genealogical link between parents increases the proportion of homozygous traits in their child.

This greater homozygosity increases the risk of being afflicted with the various recessive genetic diseases; the latter are, however, very rare; the risk, even increased, remains very slight. Above all, homozygosity causes a certain genetic impoverishment which is expressed as a very slight reduction in certain body measurements. Simultaneously, it causes a higher risk of perinatal or fetal mortality and therefore of sterility.

While it seems reasonable to state that consanguinity has undesirable genetic consequences, it seems correspondingly difficult to substantiate this qualitative statement with quantitative results. In spite of extensive and costly studies, the effect of relatedness remains ill-defined; it varies according to the populations under study and is, in any case, limited.

All things considered, it appears unlikely that the ill effects of consanguinity are at the origin of the rules which, in almost all societies, prohibit "incestuous" marriages. These effects are much too slight to be observed empirically. It is more probable that these rules are but one element in an ensemble of behaviors which insure the functioning and survival of the group. Moreover, it is noteworthy that these rules vary considerably from one culture to another: some groups encourage marriage with a maternal first cousin (the daughter of the mother's brother), others, marriage with a niece, but elsewhere these unions are considered incestuous. This has nothing to do with eugenics; rather, it is related to a system of exchange and communication which was developed to promote harmonious collaboration between clans, kinship groups, or tribes.

Mutagens in Our Environment: A Real But Exaggerated Danger

We saw that gene stability, on which the Mendelian model is based, is not absolute. It sometimes happens that an individual having received from his parents the genes a and b for a particular elementary function, transmits to his child a c gene, with a different effect. A "mutation" has taken place. This individual had received genes coding for a particular protein with a well-defined structure; the code that he passes on to his child is for a new protein. What has happened?

An intricate chain of events has to take place at the molecular level in order to achieve the production of a "gamete," that is, of a sperm or an egg: doubling of chromosomes, pairing of homologous chromosomes, construction of new strands of DNA from the previous ones, and so on. At each stage of this complex process, an error can occur; in that case, the genetic message transmitted is no longer identical with the initial message.

The nuclei of the cells, even when the cells are not being duplicated, are not completely static; they are subject to external influences that can modify the chemical structure of the molecules of DNA and, thereby, the nature of the genes which they represent.

When such modifications occur in cells other than those involved in the production of gametes, they affect only the individual in whom they occur: some of his cells carry a different genetic heritage; he constitutes what doctors call a "mosaic." Suppose that, at his conception, he received genes for blue eyes; one of the cells which, in the course of his embryonic development, participates in the formation of the iris, undergoes a mutation; a brown section appears; but the cells of the genital organs have not been changed and he transmits to his children the same "blue eye" genes that he had received. The "brown" trait is an acquired, nontransmissible trait.

The only mutations with long-term consequences are those that affect cells that lead, by successive duplications, to sperm or eggs. This is the only type of mutation that concerns us here.

The Frequency of Mutations

It is relatively easy to calculate the frequency of mutations in unicellular organisms such as bacteria. This can be done by growing a known number of cells in an environment that permits only those cells that carry a particular new mutation to grow; the mutation rate is then calculated from the proportion of surviving cells. In the case of experimental organisms that are sexual and more complex, such as the fruit fly (Drosophila), the task is less easy; it is necessary to carry out programs of cross-breeding on very large populations of flies. In the case of the human species, where experimentation is practically impossible, whether for ethical or practical reasons (related in particular to the long generation cycle), only a few very imprecise estimates are available.

Most mutations are not, in fact, expressed as soon as they occur. When the trait determined by them is recessive, many generations may pass before a homozygous individual, that is someone with two copies of the same mutated gene, exhibits the new trait. The frequency of the mutated gene cannot therefore be estimated directly; an indirect but not entirely reliable estimate can be obtained, based on more or less realistic models, and on hypotheses concerning the constancy of gene frequencies, which do not necessarily correspond to the facts. The frequencies of dominant mutations or of mutations to genes on the sex chromosome X can be calulated more accurately because these mutations are expressed in the first generation.

The rates of mutation estimated by these means vary greatly, but they are usually between 1 per 100,000 and 1 per 1,000,000. Huntington's chorea is one example of a disease whose incidence has been studied with a view to estimating the rate of mutation. This disease, which causes progressive and incurable degeneration of the nervous system, usually appears at adulthood. It is due to a dominant gene which is not expressed till the age of thirty or forty. Affected people born to parents who are free of the disease are victims of mutation. In a study carried out in Michigan on a population of 2.3 million, Reed and Neel (1959) observed 25 cases that could not be explained by heredity; 25 mutations therefore occurred in a total of 4.6 million genes, which represent a frequency of 1 per 184,000.

The frequency of mutation for a single elementary trait is therefore very low. However, when the ensemble of all traits is considered, the probability that at least one mutation may occur is relatively high. Assuming that a human gamete has, let us say, 30,000 genes, and that the average rate of mutation is 1 per 500,000 (plausible numbers, but which merely represent orders of magnitude), the probability of mutation is as high as 6 percent: of the hundreds of thousands or millions of gametes produced by an individual, a very significant number are therefore carriers of mutations.

Natural Mutagenic Agents

These mutations occur spontaneously, at random, and the underlying mechanisms are not yet understood. However, it is possible to induce them by certain "agents" called "mutagens," such as radiation and chemicals.

As early as 1927, the effect of radiation was illustrated by H. Muller: exposure of *Drosophila* to increasing doses of X-rays brings about a proportional increase in the frequency of mutations. When these rays (and the various other forms of ionizing radiation such as α rays, β rays, γ rays, ultraviolet rays, neutrons, etc.) penetrate the body's tissues, certain electrons are ejected from their normal paths; the molecules "unhinged" by the loss of these electrons become very reactive; they may in particular combine with the DNA bases that carry the genetic information, and cause this information to be changed.

It quickly became obvious that, at least in *Drosophila*, the natural radiation (cosmic rays, terrestrial radiation, radioactive atoms in food), to which all living organisms are exposed, could not be the cause of all spontaneous mutations; it would need to be a thousand times more intense than it actually is to explain the rate observed. There are other factors involved, chiefly certain kinds of chemical reactions. Each cell resembles a chemical factory with a wide variety of products; of the molecules that are produced, some, such as nitrous acid, may interact directly with the constituent bases of the DNA; others, such as 5-bromouracil, have a chemical structure that is close to one of these bases and may be substituted for it. In each case, the genetic information is altered, a mutation occurs.

The degree to which spontaneous mutations are due to natural radiation and the degree to which they are due to chemical agents may not be the same in *Drosophila* as in humans. The organisms involved in each case are very different; the human organism is much more complex and has a much longer generation span. Obviously, there is no question of experimenting with the human species the way we do with flies; we can, however, study a mammal that is closer to us, the mouse. It seems that, in this species as in *Drosophila*, natural radiation causes but a very small fraction of spontaneous mutations, about 1 percent.

An estimate of this percentage in the human species can be calculated by comparing the dose of radiation received naturally with the dose required to double the normal rate of mutation. These doses are generally calculated according to the amount of energy absorbed and the basic unit is called a "rad" (1 rad is the equivalent of 100 ergs per gram of tissue). It is assumed that the level of radiation absorbed during the thirty years from birth to adulthood by the reproductive organs is about 3 rad (half of which is due to terrestrial radiation and a quarter to cosmic rays). However, this estimate is imprecise because it depends on the altitude and especially on the nature of the terrain (this dose is considerably higher in granitic regions; in one region in the state of Kerala in India, it is six times higher than average because of the presence of monoazite sands, rich in radioactive atoms).

The estimated dose required to double the rate of mutation is even more imprecise. It is probably somewhere between 40 and 200 rad; these figures would indicate that natural radiation causes between 7 and 1.5 percent of spontaneous mutations. In spite of its imprecision, this estimate will be useful to us in predicting the possible genetic consequences of the increased radiation to which modern man is exposed.

In conclusion, it is to be noted that the temperature of the reproductive cells plays a definite role in the frequency of mutations in *Drosophila;* this frequency increases when the temperature rises. It is probable that the same is true of our species. The American geneticist Curt Stern (1960) pointed out that the temperature of the genital organs of fully-clothed men is more than 3° higher than that of naked men: in prohibiting nakedness, civilized man has therefore increased the rate of mutation.

Artificial Radiation

The development of technology has considerably increased the dose of radiation to which we are exposed. Above a certain level, this dose becomes lethal: there is almost no chance of survival when it is greater than 450 rad; above 100 rad it is associated with major disorders such as nausea, vomiting, anemia, and, above all, a weakening of the body's defense mechanisms against infection.

These are, however, individual consequences and, serious though they undoubledly are, we are not concerned with them here: we are interested in finding out how future generations will be affected, that is, how radiation affects the genes that are transmitted by an individual, irrespective of that individual's own state of health.

The most spectacular change in our environment, with regard to radiation is, of course, the use of nuclear energy in its various forms, both peaceful and military, but the most important, qualitatively, is the invention of X-rays. Their efficiency in medical diagnosis sometimes led to an excessive use of these rays. It is important to remember, however, that it is only those rays that reach the reproductive organs, the ovaries and testes, that affect the rate of mutation. On average, in developed countries, the total dose received by these organs is about 1.3 rad for men and 0.3 rad for women, whose ovaries are better protected. These amounts (which may, of course, be much greater in the case of people requiring special treatment) indicate that the consequences for the genetic heritage are limited. The use of X-rays increases the amount of radiation received during the course of a "normal" lifetime by about one third, and we saw that this radiation causes only a small fraction (between 7 and 1.5 percent) of spontaneous mutations.

It may come as a shock to some to hear that the use of nuclear energy, in its present state of development, has an even smaller effect on the genetic heritage.

Humanity's collective conscience was traumatized by the extent of the destruction which the atomic bomb makes possible; the atomic dawns of Hiroshima and Nagasaki nonetheless killed fewer people than the phosphorescent nights of Dresden or Tokyo; it is the disproportion between the means, a single bomb weighing a few hundred kilograms, and the result achieved, a city of several hundred thousand razed in a few seconds, that generates the feeling of being in the presence of a new and terrifying power.

In a small carefully-raked gravel courtyard that can be visited by tourists to Los Alamos (New Mexico), near the factory where the first atomic bombs were manufactured between 1942 and 1945, are displayed two life-sized replicas of "Fat John" and "Little Boy," the two bombs that so brilliantly inaugurated the atomic era. One needs only a little imagination to be disturbed by the diabolic power enclosed in these modest white containers, so seemingly harmless. A demon has been let loose and, ever since, everyone has been trying to enlist his services.

However, to combat this demon effectively, it is important to choose appropriate weapons; attributing imaginary faults to him, fighting him on ground that is not unfavorable to him, can only strengthen his position in the long run. Let us not use unfounded arguments against him.

Within the very narrow perspective which we are adopting here, the danger represented by the atomic bomb seems slight compared to the fears it has inspired. The various A and H bomb tests carried out in the atmosphere, since the last world war, by the members of the "atomic club" have released radioactive products which are now revolving in the upper atmosphere and slowly leaking back down to earth; the radioactive power of some of these products, such as strontium-89, lasts for only a very short time (50 days); in others, on the contrary, it lasts for centuries: the half-life of carbon-14 lasts for more than 5,700 years. Since some radioactive debris remain potent for such a long period of time, the cumulative effects of testing could lead to dangerous situations. Bear in mind the following overall estimate of the amount of radioactive debris in the atmosphere: all the explosions that occurred up to 1970 will cause an average increase of 0.24 rad (Tobias).

This level can be considered either high or low: to those who were anticipating disastrous genetic contamination as a result of these experiments, it shows that, in fact, our environment has not really been changed; the increase in the level of radiation to which each individual is exposed is very much lower than the differences noted from one region to another, depending on soil composition. The harm has not yet been done.

However, this increase, though slight, is not insignificant; it results from a small number of experiments. If this irresponsible policy had been continued or if it were ever to be resumed in the future, it could, because of the accumulation of radioactive debris, lead to a much more disturbing situation.

Fear of this cumulative effect is the source of concern about the development of nuclear power plants: in spite of all precautions, a tiny fraction of radioactivity may escape into the gas emanating from chimneys or into the cooling water. Precise measures of these dangers are not easily obtained. The reputable *Encyclopaedia Britannica* (Tobias) gives the following estimate in its 1977 edition: if the current trend toward the development of these plants continues, the natural dose of radiation will be doubled before the end of the century by the various artificial sources. If even this level is considered acceptable, it is clear that current policies will have to be changed; otherwise, the level will quickly be tripled and quadrupled.

It is important to realize that neither the use of coal plants, nor even that of atomic fusion (if its industrial use should one day become feasible), would entirely solve the problem of radioactive pollution. The energy race itself is at the root of this problem.

Mutagenicity of Chemical Products

Chemists tend to have especially fertile imaginations. Our day-to-day life has been profoundly changed by their various productions. However, we saw that our own cells are themselves chemical factories. Some of the things that enter these factories, or that are made there, may react with the DNA bases, modify the structure of our chromosomes, and induce mutations. How can we know whether a particular product, be it natural or artificial, has this terrible power? The truth is that we are almost totally ignorant in this domain.

The mutagenicity of most food additives is tested through experiments on cell cultures or on lower organisms; but the conclusions drawn from these experiments are not necessarily applicable to humans: a particular product that causes a mutation in bacteria may be harmless for us if it is unable to penetrate to the nuclei of our cells, or if it is destroyed by certain reactions; another product that is harmless for those species tested in laboratories may be active in human cells (such as the infamous thalidomide, which was responsible for the birth of so many children without arms and legs).

Direct experiments on humans are, of course, out of the question for ethical reasons and also because no firm conclusions could be drawn from them. In the event that a particular product increased the rate of mutation by, let us say, 20 percent, one would have to use hundreds of men and women as guinea pigs and wait several generations before detecting a significant effect. (Even in the case of thalidomide, which caused especially spectacular defects, it was several years before the cause of this "epidemic" was discovered.)

There is no foreseeable remedy for our inability to detect the long-term mutagenic effect of the new chemical substances which we sometimes use in large quantities and this is especially serious. A small fraction, less than 7 or 8 percent, of spontaneous mutations is caused by ionizing radiation; the remainder are linked essentially to the various chemical processes occurring in the cells. Any modification of the conditions under which these processes occur could therefore have dramatic consequences. This, however, is no more than a possibility; most substances do not, in all probability, harm our genetic heritage. This, at least, is our hope, but we can by no means prove it objectively. This lack of information makes prudence advisable, but our society seems uninclined to adopt this attitude.

FOUR

Human Races: An Ill-Defined Concept

THE MULTIPLICITY and complexity of humanity inspires the need to divide it into categories and groups, by assigning those individuals who resemble each other most to the same category. For this classification to have biological significance, those traits which allow one to see similarities must, of course, be hereditary and must also be to some extent stable from one generation to the next.

Initially, classifications could only be based on directly observable traits, the colors and shapes of individuals, for instance. Such classifications were often subtle and took account of complex parameters, but they were, of necessity, limited to the "realm of phenotypes." Taxonomists defined various races according to skin colors (black, white, or yellow), hair texture (curly or straight), the relationship between the length and breadth of the skull (dolichocephalic or brachycephalic), etc. Depending on the traits studied, the classes or "races" identified were different, and there were lively polemics between those who detected 4 principal races and 25 secondary ones and those who found 20, or 29, or 40 races.

The discoveries of genetics made it possible to outline the problem more precisely by giving a more objective basis to the concept of race: a race is a group of individuals who have a large part of their genetic heritage in common. This time, the classification is based on intrinsic characteristics of the various human groups, independent of the conditions under which they live; it concerns the "realm of genotypes." Therefore, conclusions sufficiently clear to win general agreement can reasonably be hoped for.

Unfortunately, scientists failed to discard the old concepts when interpreting the new observations, or, to use a biblical metaphor, they put the new wine into the old casks; in spite of the remarkable progress of knowledge, confusion on this point has merely worsened. Those biologists who have had the courage to go against generally accepted ideas (J. Hiernaux 1969; J. Ruffié 1976; or A. Langaney 1977, for instance, recently in France) have not received sufficient attention; public opinion is still marked by theories that are completely outdated but nonetheless have the authority of ancient myth.

Race and Racism

To begin, it is useful to compare the two terms, race and racism:

—the first is the object of legitimate scientific inquiry, based on objective facts: the aim is to develop methods of classifying individuals with a view to defining human groups, "races," that are relatively homogeneous;
—the second evokes an attitude of mind, which, of necessity, is subjective: the different races are compared and evaluated with a view to establishing a hierarchy.

These are, of course, two distinct activities: it is possible to try to define races without being in the least "racist" in the sense that we have indicated. It is to be noted, however, that this is, in practice, only a theoretical possibility. The need to define races is rarely motivated by the taxonomist's single-minded concern with ordering all his data; it comes from the desire, so highly developed in our society, to differentiate the group to which we ourselves belong from other groups. It corresponds to the Platonic idea of "type." We can define the human species, but it is difficult to outline in any kind of detail the ideal human type; several types have to be considered: white, black, Indian, Eskimo, etc.

Without claiming that all responsible taxonomists fell to such excesses, we quote the following extracts from Crozat's *Géographie universelle*, published in 1827, just a century and a half ago, to illustrate to the point of caricature, where this notion of "type" can lead and what confusion it can create:

> The Chinese have a wide forehead, a square face, a short nose, large ears and black hair . . . They are naturally gentle and patient but egoistic, proud . . . Negroes are generally well-built and robust, but lazy, deceitful, drunken, gluttonous and slovenly . . .
> The inhabitants of America are agile and swift; most of them are lazy and indolent, some are very cruel . . .

This is quite enough of such nonsense, which, it must be remembered, was written, not by a novelist expressing his personal opinions, but by a geographer concerned with producing scientific work. The point most clearly illustrated by these quotations is that a classification is usually based on a variety of criteria, some of which are objective, others subjective, and that it rarely avoids establishing a hierarchy: races are different, therefore some are "better" than others. The extremes to which some dictators have carried this notion are all too well-known.

They are merely translating the ideas of certain scientists into politics and action. Since Darwin, our perception of the progressive transformation of living things, plants, animals, or humans, is based on such concepts as the struggle for existence, the survival of the fittest, the elimination of the unfit, and the propagation, over many generations, of favorable traits. These concepts, which were initially developed with individuals in mind, have been extended, almost unchanged, to groups of individuals, or races. Differences between races have come to be seen as the products of varying degrees of evolutionary development and to be thus perceived as inequalities. Even in casual conversation, the average person tends to make remarks which imply that racial inequalities are obvious: some races are superior (usually that of the speaker), others are inferior.

Of course, most French people claim sincerely that they are not racist; South Africans, North Americans, Germans, or Russians are terribly racist, but not us. It is just that we rightly feel superior to Arabs, blacks, gypsies, or Indians, not to mention various other groups who are less well-endowed than us by nature and "who are not like us," if you see what I mean. Let's face it, racism, that is the feeling of belonging to a biologically superior human group, is almost universal.

Examples of this unconscious racism abound. The one that surprised us most was undoubtedly this unexpected sentence taken from the chapter on the duties of a colonel in the *Réglement du service dans l'armée:* "The colonel . . . indicates the most appropriate means of developing patriotism: fortifying love of country and a sense of racial superiority . . ." This regulation was not directed at the German army in the Nazi period but at the French army, and the document was printed in 1957. That a sentence like this was approved by several government ministers and army chiefs-of-staff proves how natural it seems to most people to define a "French race" and to glorify it relative to other races.

A scientist who discovers that the data available to him do indeed confirm racial inequalities and therefore constitute a potential basis for a racial hierarchy should not hide this conclusion. Respect for truth is the ethic of science. In the opposite event, however, he must not hesitate to proclaim this truth and, when necessary, to combat entrenched ideas even when they are virtually unanimously accepted. It is important at this point to sum up: what does science, especially genetics, have to say about the concept of race?

What Does Classification Involve?

Defining races involves classifying the teeming masses formed by the few billion people alive at present and their several billion ancestors.

Similarly, the definition of species implies that all living organisms must be divided into groups. However, in the latter case, we have a rather specific criterion for deciding whether two organisms belong to the same species or not: their capacity (real or potential) for interbreeding. Some borderline cases are, of course, difficult to decide, but the notion of belonging to the human species is, on the whole, clear and comprehensible; all individuals belonging to this species, no matter how far apart in space and type, whether Australian aboriginals, Northern Eskimos, inhabitants of Tierra del Fuego, or Melanesians, are potentially interfertile.

There is, however, no such criterion for deciding whether two individuals belong to the same race or not. Nevertheless, we regularly make this kind of decision without hesitation; we need no special knowledge to decide almost unerringly whether a particular man passing by us on the street is Chinese or Arab or Indian. We are now going to examine the mental process that leads us to make this kind of classification.

We are aware of the vast number of different individuals belonging to our species; these individuals are much too numerous for our intellect to be able to compare them easily with each other; we therefore replace them with a set of categories, much smaller in number, in such a way that each individual belongs to one category and one only, and that all individuals within the same category are "similar."

Therefore, we are first of all obliged to specify what we mean by "similarity"; more precisely, we must indicate what criteria are being taken into consideration; if there is just one criterion, skin color on the inside of the arm, for instance, it is easy to measure the similarities. If, however, there are more than one, for example, this color and head size, we have to arbitrarily define a global measure that includes both these parameters simultaneously. A technique for defining this measure has been developed by mathematicians; it involves calculating a "distance": the smaller the distance between two individuals, the more similar they are. There are several formulae for making this calculation: depending on whether we are using "Euclidian distance" or "Manhattan distance" or "chi squared distance," a single set of data will yield several different sets of distances. The choice is practically infinite, so rich is the imagination of mathematicians.

Suppose that, having chosen certain criteria for a classification and a formula for calculating distances, we managed to determine all the distances d_{ij} between each individual i and each of the others j (for the 4 billion or so people alive at present, the number of distances will be in the order of 8 billion billions). The categories that we are trying to establish will be meaningful only if the distances between individuals within the same category are, at least on average, significantly smaller than between individuals in different categories. Here again a large number of methods for achieving this have been developed and each one leads to a different result.

The simplest method, which no doubt is also the closest to the intuitive, natural approach, involves the construction of a "tree": the two closest, or most similar, elements are first of all put together to constitute a class made up of just these two elements, then the closest classes are put together; in this way, the number of classes is gradually reduced until all that remains is one class that includes all the elements.

To illustrate this process, consider a very simple example: a species consisting of only 10 individuals a, b, . . . , j. Let us suppose that we have chosen the classification criteria; we have obtained the measure of each criterion for each individual; finally, we chose a "distance" formula that enabled us to calculate the 45 numbers $d(a, b)$, $d(a, c)$ . . . , $d(i, j)$, or distances, characterizing the degree of dissimilarity between individuals. We notice that the shortest distance is $d(a, f)$; we therefore group a and f together in a "class" called K_1, which we now consider to be a fictitious "individual." Then, we calculate the 36 distances between K_1, b, c, d, e, g, h, i, j; this time, we notice that $d(K_1, c)$ is the shortest of these distances so we put c with the class $K_1(a, f)$ to constitute class K_2 and so on. To do this, we had to decide on a method for calculating the distance between a class, such as K_1, and an element, and for calculating the distance between two classes, which can be done in many different ways.

Eventually, having made all these arbitrary decisions, we get a tree similar to that illustrated in figure 5. How can it be used to define races? One further choice remains to be made, that of the

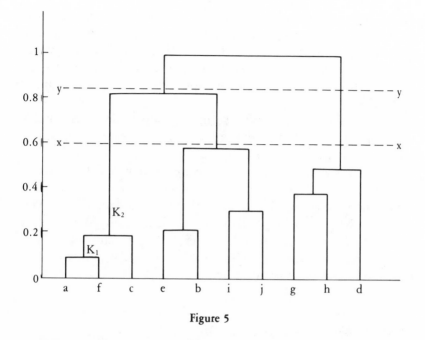

Figure 5

number of races to be defined, a number which, of necessity, is between 1 (race and species are thus one and the same thing) and 10 (as many races as individuals, which makes our efforts meaningless). If we wish to identify 3 races, we will cut the tree at the level of the xx' line, which gives us the "races" (a f c), (e b i j), and (g h d); but if we prefer just two races, we must draw the line yy' which isolates the "race" (a f c e b i j) from the "race" (g h d); etc.

It is to be noted, however, that the level at which we drew the horizontal lines to group individuals or classes has a specific meaning: it represents the loss of information that occurs when the initial data concerning individuals are replaced by global data concerning classes. We see that, in our example, grouping in 3 races involves a loss of only 60 percent of this information (the line xx' which cuts the tree where it has three branches is indeed at the height 0.6), while grouping in two races involves a loss of more than 80 percent. In order to lose no information, one has to stay at the zero level, which amounts to making no grouping; on the contrary, in order to group all ten individuals into a single category, all the information must be lost. Using the diagram, we can thus make an informed decision regarding the choice of the level at which to cut our tree, and therefore of the appropriate number of races.

This rapid explanation will seem very superficial to researchers accustomed to this kind of work. Over the past twenty years, techniques for analyzing data, for extracting a thread of meaning from a jumble of overabundant facts, have been developed and refined to a remarkable degree. However, our intention was not to give a comprehensive lecture but to show that the business of classifying, which may seem so simple and natural, is in reality very complex and that the outcome depends on very arbitrary choices. It is not that the result of a classification is of no value, it is just that one needs to be aware of its relativity.

"Phylogenetic Trees"

The mechanism involved in classifying, which we have just analyzed in the form of the construction of a tree, can be applied to any collection of objects whatsoever, whether it be the stock of a hardware store, the various languages spoken on our planet, the animals in a forest, or the individuals in our species. However, in the latter case, the aim is not only the grouping of similar individuals into relatively homogeneous classes, it is to find a set of historical facts: their genealogies, the complete pattern of their ancestry, generation after generation.

Two individuals with common ancestors have received identical genes from those ancestors. This genotypic similarity will be reflected to some extent at the phenotypic level. When we make groupings based on comparisons between phenotypes, it is reasonable to hope that the closer two individuals are, the more common ancestors they have; by making a tree diagram, we will obtain a rough outline of their patterns of relationship and descent; we will draw what is called a "phylogenetic tree." Consider a population that has undergone successive divisions during a process of fission similar to that outlined in figure 6. Each group, after a certain period of autonomy, divides into two populations which remain completely and permanently separate and which, in turn, later undergo similar division.

An individual belonging to group A in our diagram has more genes in common with an individual from group B than with an individual from C or G, because in the case of A and B, one finds common ancestors after going back a fewer number of generations. A study of resemblances between present-day populations can therefore be attempted with a view to reconstituting the history of their paths of descent. It is easy to see why this information is of such interest to historians and ethnologists who are eager to learn the origins of the people that they study.

This work has been accomplished with remarkable precision for the ensemble of species that constitute the living world, with each species being considered as a single homogeneous group. We now have trees displaying the whale as well as the fly, the human being as well as the trout, and also their distant common ancestors. Figure 7 shows a plausible tree for the appearance of various animal species, established by American geneticist D. Hartl (1977) based on structural differences between the various proteins common to all species. This tree is very similar to those that had been established by taxonomists based on anatomical comparisons. This re-

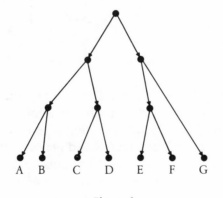

Figure 6

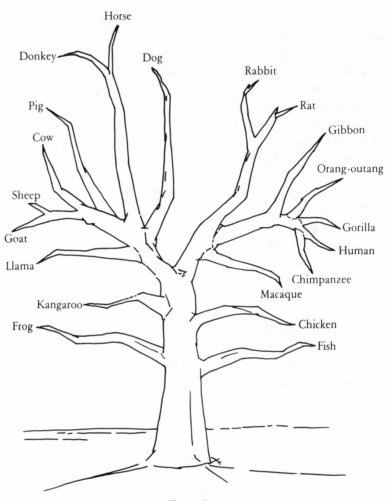

Figure 7

construction is facilitated by the fact that the various species satisfy the condition that we hypothesized for the drawing of a phylogenetic tree: populations are subject to divisions, but not to fusions; once separated, they remain so permanently. When, after chromosomal rearrangements or accumulated mutations, a new species appears, cross-fertilization with the original species is impossible (or, as in the case of the mule, the progeny are sterile, but this sterility has the same consequences), and the genetic separation is complete.

On the contrary, when a population is split into two groups which remain interfertile and therefore still belong to the same species, though they evolve separately and gradually become two distinct "races," genetic exchanges remain possible between them, either due to migration or to the complete fusion of the two temporarily separate groups. This type of evolution cannot be represented by a tree of the kind shown in figure 6. It is, rather, a complex network, such as that represented in figure 8. This network cannot in any way be compared to a classification tree. Even the highly sophisticated mathematical techniques that made it possible to construct these trees are completely incapable of reconstructing networks that have been complicated by fusion between groups.

We will see that, in spite of this fundamental impossibility, many researchers have tried to use data gathered from populations alive today as a key to possible historical links between these populations. These attempts are not useless so long as their limitations are borne in mind. Any results derived from them must be considered merely as a confirmation of other data obtained from independent sources. In any case, the aim is usually not to study the phylogeny of humanity as a whole, but simply to analyze the relationships between the various groups living in a limited geographic area.

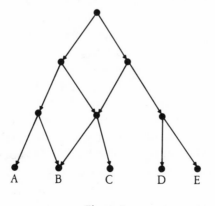

Figure 8

The Question of Skin Color

In defining races, the trait that is spontaneously brought to mind is that which is most easily observed: skin color. This trait is obviously hereditary, but the underlying genetic determinism is not fully understood.

The first point to be noted is that, contrary to widely-held opinion, the pigment melanin, which is responsible for skin color, is present in the epidermis of white-, yellow- and blackskinned people but in very different doses. The differences are therefore quantitative and not qualitative. There is considerable variation within each group and the difference between two people within a population can be much greater than the difference between the averages of two groups belonging to distinct "races." In a recent study entitled "La quadrature des races," André Langaney (1977) shows that it is possible to go without discontinuity from those people with the palest skin (Northern Europeans) to the darkest (the Saras of Tchad) with intermediate types being chosen from only two other populations (North Africans and Bochiman Bushmen of the Kalahari desert).

Studies on crosses between whites and blacks and between their descendants have shown that this trait follows a typically Mendelian pattern. It behaves as though it were governed by 4 pairs of genes with complementary effects; the actual mechanism is not doubt far more complex, but this simple model accounts very well for what is observed. Assume then that "whites" have eight w genes for white skin and "blacks" eight b genes for black skin. All the intermediate shades are possible, depending on the distribution of ws and bs within those eight genes.

Studies done on American blacks confirm this genetic model. This very heterogeneous group consists of all North Americans who have among their ancestors Africans deported as slaves between the beginning of the seventeenth century and the middle of the nineteenth century; they also actually have a considerable number of Europeans among their ancestors, since young female slaves frequently bore children that were fathered by their white masters. In fact, comparisons between the frequencies of certain genes in

African populations in the Bénin Gulf region (which were the chief source of slaves), in the Anglo-Saxon populations of Europe and in American blacks, have led to the conclusion that 25 percent of the genes in the latter group are "white."

For instance, in the case of the Rhesus blood system, the frequency of a certain gene called R_0 is 63 percent for the Africans while it is only 3 percent for the Europeans. For American blacks, an intermediate frequency of 45 percent is found, which is consistent with the hypothesis that a quarter of their genes are European. This is, of course, only a general estimate; the actual proportions are no doubt very different according to regions and families. (The genetic input of whites is much greater among American blacks in the North and West than in the South.)

According to this hypothesis, each of the genes controlling skin color in an American black has 1 chance in 4 of being a w gene. The probability that all the 8 genes involved will be w is therefore equal to $(1/4)^8$, or about $1/65{,}000$. In other words, of the 20 million or so American blacks, several hundred have only genes for white skin and are actually white. Similarly, the probability of having 8 b genes is $(3/4)^8$, or about $1/10$: there are, therefore, only 2 million "American blacks" whose genes for skin color have come exclusively from the founder African population and who are as black as their ancestors. This calculation can be continued and the distribution of "blacks" according to the number of b genes (between 0 and 8) which they carry can be estimated. This distribution turns out to be very similar to the actual distribution of skin colors in this population, which proves that the "model with four pairs of genes" is a good representation of reality.

Populations with very dark skin are found mostly in Melanesia (i.e., the group of islands situated in the Southwest Pacific), in the Indian peninsula, and in Africa south of the Sahara, regions that are all close to the equator. This fact is used in support of the theory that black skin has a greater adaptive value in hot countries. Later, we will see that even this idea, which is so widely accepted, can be disputed. For the present, we merely point out that these three groups of populations do not, in any sense, constitute a "race"; apart from skin color, they are different in every respect: analysis of their blood systems shows, for example, that they cannot be considered to be offshoots from one initial group; their "phylogenetic tree" cannot be represented as a single trunk with three branches. If this were so, traits other than black skin would have been transmitted from the ancestor population to all three groups, but this does not seem to be the case. This finding shows that no classification based on skin color alone can have biological significance; it is very inconvenient for those who imagine that a definition of races can be based exclusively on this criterion (which, of course, anthropologists no longer do); but there is no getting away from it.

Lastly, we find that even though skin color is the most obvious trait, and the easiest to compare, it involves only a tiny fraction of our genetic heritage (no doubt 8 or 10 genes out of several ten thousand); no other major biological trait seems to be linked to it; it cannot therefore be used as a basis for classifying populations in a meaningful way: how many individual and collective tragedies could have been avoided, and could still be avoided, if this simple fact had been, or were at last, accepted by everyone.

Are there not other physical traits, relatively easy to measure, which could be substituted for skin color as a basis for classification?

Height, head length and breadth, the relationship between the last two (the cephalic index being the basis for differentiating between "brachycephalics" and "dolichocephalics") and many other body measurements can be used to determine similarity and dissimilarity between individuals or between groups. However, the genetic determinism of those traits is far from clear, and, in the case of most of them, it is completely unknown. Given our present state of knowledge, it is impossible, and probably will be for a long time to come, to infer genotypes from information gathered on phenotypes.

Moreover, some of these traits, height for instance, are not at all stable, in spite of being genetically determined. In all industrialized countries since the beginning of this century, there has been an extraordinarily rapid increase in people's height. According to G. Olivier's (1977) recent study, the height of French conscripts at the age of 20 was:

| 165.4 cm in 1880 | 165.8 cm in 1900 | 165.7 cm in 1920 |
| 168.5 cm in 1940 | 170.0 cm in 1960 | 172.3 cm in 1974 |

These data indicate that the increase is occurring more and more rapidly. It cannot be caused by genetic changes; only environmental influences (but which ones? one can do no more than guess) could have produced these changes in such a short time. The fact that height varies to this extent means that all hope of using it to compare different populations or to reconstruct their "phylogenetic tree" must be abandoned.

In confining itself to quantitative traits, the genetic significance of which, as we shall see in chapter 6, is difficult to interpret, anthropology risked becoming stuck in an impasse; advances in biochemistry opportunely supplied it with facts that made a new stage possible: these facts pertain to traits, essentially the blood systems, with a genetic determinism that is so clear-cut as to make the inference of genotype from phenotype much easier.

Blood and Its "Systems"

The first blood-group "system" was discovered in 1900 (the same year that, by pure coincidence, the rediscovery of Mendel's laws heralded the development of modern genetics). The Austrian biologist Karl Landsteiner noticed that certain people's blood is apt to agglutinate that of certain others, which explains why accidents sometimes occur during blood transfusions. He was thus able to identify four "groups": A, B, AB, and O. Analysis of the transmission of this trait in families showed that, in each individual, it is governed by one pair of genes; for each of these genes, there are three possibilities: *A, B* or *O;* moreover, the *O* gene is recessive to the *A* or *B* genes. Therefore, the correspondence between the pair of genes carried (the genotype) and the trait manifested (the phenotype) is:

Genotype	Phenotype
AA and AO	A
BB and BO	B
AB	AB
OO	O

Twenty-seven years later, Landsteiner discovered a second system, called "MN," and, in 1940, a third, the well-known Rhesus system. Since the last war, discoveries have continued at an ever increasing rate. The study of the structure of hemoglobins, as well as of the properties of red blood cells, of white blood cells, and of serum, has led to the identification of more than seventy systems and the list is increasing every year.

The big advantage of these traits is that they give us information on genotypes. Even though this information concerns only a tiny number of genes and is therefore very limited, it allows us to compare populations, based on some objective measures that are independent of the effects of environment. A person who is born with an A and a B gene belongs to the group AB, regardless of whether he is young or old, starving or well-nourished, living in a tropical forest or in the Canadian North. To classify populations, all that is required is that sufficient data be accumulated, based on blood samples from the various human populations. This work has been undertaken by numerous teams who have left scarcely any "unknown territories" on the maps which chart their findings; Professor A. E. Mourant's (1976) atlas is proof of this. However, the samples from many regions are far from being representative and the results are imprecise. The task is, therefore, not completed and much work still remains to be done.

Nonetheless, some lessons can be learned from the data currently available, in spite of their incompleteness and imperfection.

The first concerns the rareness of "marker" genes: a "marker" gene g is one which is found in a population P but in no other. It is, therefore, a specific trait which differentiates P: an individual with this g gene can belong only to population P. Note, however, that the converse is not true: *all* individuals in population P do not have the g gene and it can even be relatively rare in this population. Whether it was introduced from outside by a migrant or is the product of a mutation within, it is not necessarily very widespread. In spite of extensive research, relatively few "markers" have been found. The clearest example is that of a certain gene a for the system "Diégo" which was discovered in Venezuela in 1954. This gene, which reaches a frequency of 40 percent among certain

South American Indian tribes, is completely absent in Central Africa, as it is among Polynesians, Papuans, and Australian aboriginals; in Europe, it is found only in rare cases; on the other hand, it is quite widely represented among most of the populations of Far East Asia.

Similarly, some genes associated with the Gm system, which we will be discussing in detail later, can be considered specific to Central Africa, for instance, *GmG* and *GmH*. Though it is not completely impossible to find one of these genes elsewhere, it is extremely rare.

In the case of all other genes, regardless of the system involved, no specificity has been found. What distinguishes two populations is not the fact of having or not having a particular gene, but the fact that the frequencies of this gene are different. The criterion is not one of "all or nothing" but, rather, "more or less."

To compare populations, we must therefore synthesize into a single criterion their greater or lesser similarity, which is determined by the degree of similarity between the frequencies of various genes within them. Consider an imaginary example, that of 4 populations in which the frequencies of 4 genes, a_1, a_2, a_3, a_4, for a certain blood system are known. The frequencies, expressed as percentages, of each of these genes are indicated in table 1:

Table 1

Gene→ Population	a_1	a_2	a_3	a_4
I	2	3	75	20
II	1	49	20	30
III	40	30	3	27
IV	27	30	40	3

The problem is how to decide which populations are most similar and which are most different. The difficulty becomes obvious when we try to compare the data: I resembles II for genes a_1 and a_4 but differs from it for a_2 and a_3; III resembles IV for a_1 and a_2 but differs for a_3 and a_4 and so on. To reach a decision, we need to calculate a distance, that is, a number that increases as the populations become on average more dissimilar. We saw that there are many different formulae available and that they can produce widely different results; many population geneticists, especially British and Americans, use a distance called the "arc cosine" in such cases. Without going into technical details which are of little interest here, we can say that, for our example, the result is as follows:

$$d \text{ (I-II)} = 1 \qquad d \text{ (I-III)} = 1.71 \qquad d \text{ (I-IV)} = 0.86$$
$$d \text{ (II-III)} = 1.05 \qquad d \text{ (II-IV)} = 0.98 \qquad d \text{ (III-IV)} = 0.76$$

the distance between I and II having been arbitrarily assigned a value of one. It is useful to graph these results in such a way that the populations are displayed as points separated by distances that are as similar as possible to the distances between populations. Here, we get 4 points like those on figure 9. The data available on many blood groups in a large number of populations thus allow us to calculate a set of distances and to draw genetic maps which sometimes show a surprising divergence between genetic distances and geographical distances.

Ph. Lefevre-Witier's work (1974) on the populations of North and West Africa is a good example of this kind of study: having isolated 26 "populations," he compared the frequencies of the various genes associated with 5 blood systems. The 325 pairwise dis-

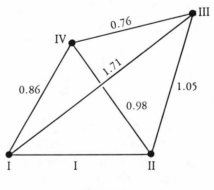

Figure 9

tances between these populations were used as a basis for the map featured in figure 10. From this map, we can see that the Kel Kummer Tuareg of Mali, the R'Gueibat of Algeria, and the control group of French from the eastern Pyrenees are very similar; at the other extreme of the distribution, there is a group consisting of the Gagou of Ivory Coast and the "Iklan," descendants of slaves orignially from the Bénin Gulf and still living in the Tuareg tribes;

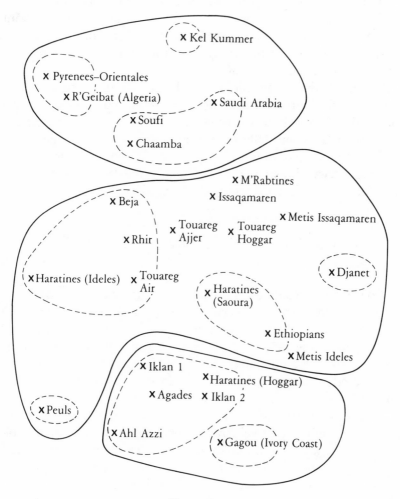

Figure 10

between these two extreme groups, there are various Saharan tribes, the Peuls, and an Ethiopian group.

This kind of map is enlightening and very useful for research, but it is easy to overestimate its significance. First of all, the samples used to represent the populations are often very small; in the case of small groups such as the Kel Kummer Tuareg or the Iklan, the sampling is sufficient to insure a good representation of the entire group, but certainly not in the case of populations that are as widespread and diverse as the Peuls or the Ethiopians. The points on the graph do not really represent these populations but rather particular samples of them; other samples might be situated on a completely different zone of our map.

Similarly, the graph might have been very different if data on other blood groups had been used. We objected that skin color is determined by only a few genes and is therefore not representative. We must avoid falling into the same trap with hematological data. The possible differences according to the blood groups considered are well illustrated by a comparison between the three best documented systems: Rhesus, Gm, and HL-A.

Genetic Systems with a High Degree of Polymorphism: Rhesus, Gm, HL-A

Collection of data on some systems, though extensive and worldwide, has led to the discovery of only a small number of different genes: only 3 genes have been found for the Duffy system which was discovered in 1950. For others, on the contrary, numerous genes were identified quite rapidly and new ones are constantly being discovered. These multi-gene systems are said to be "polymorphic."

The Rhesus system belongs to this category: the well-known "positive" and "negative" traits were quickly seen to be just one aspect of a whole so complex that the debate about the genetic mechanism involved is still not closed. Note that, up to the present, more than 20 different genes have been listed.

The Gm system, discovered in 1956, is under study in many

laboratories, for instance that of Claude Ropartz in Rouen (1971). This system is not a characteristic of the red blood cells but of the proteins in the serum, the immunoglobulins which recognize "foreign" substances and neutralize them. Some of these proteins, the IgGs, have variable structures: the study of their transmission within families led to the identification of 12 different genes (from *A* to *L*); this list is, of course, provisional.

The HL-A system has been studied very intensively because of its involvement in the rejection of tissue transplants. Since 1958, numerous teams, in particular that of Jean Dausset at Saint-Louis Hospital (1973), have successfully coordinated international studies which have led to the discovery of the underlying genetic mechanism: it is now generally accepted that 4 pairs of genes are involved. These genes are located very close together on chromosome number 6. For the first site, 20 genes have been identified, for the second 30, for the third 6, for the fourth 11; these numbers increase every year.

The remarkable richness of these systems makes them useful tools when it comes to comparing populations. However, this very richness makes their use more complicated: a very small sample suffices to tell us the frequency of the genes for the Duffy system in a particular population; but a huge sample is required if one wants to estimate the frequencies of the various Gm or HL-A genes, some of which may be extremely rare. In order to be able, without too great a risk of error, to assume that a particular gene is absent, and such assumptions are often crucial, samples from a large fraction of the population must be studied.

Without going into details, it should be noted, in agreement with A. Langaney (1977), that the information made available to us by these three systems on the colonization of the earth by human populations, and on their subsequent differentiation, is largely contradictory:

— the *r* gene for the Rhesus system is very rare in Oceania and the Far East, common in Africa, in India, in the Middle East, and above all in Europe (its frequency is at more that 50 percent among the Basques and the Bedouin of Sinai);

— the R_o gene, which seems to be due to a recombination that occurred somewhat later in human evolution, is at a high frequency in black Africa only.

Based on this single system, the tree for the three big groups would take the shape shown in diagram 1.

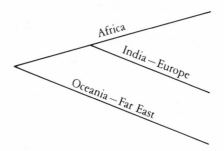

If the Gm system is used as a basis for the classification the result is quite different:

— the *A* gene, the most frequent in Europe, is widespread in Asia and all of the Pacific; it is unknown in black Africa;
— the *G* and *H* genes which are common in black Africa are practically absent from the rest of the world;
— the frequencies of the various genes are very different in the Far East and in the India-Iran zone;

The tree based on the Gm system looks something like what is shown in diagram 2.

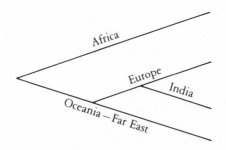

The HL-A system involves so many genes that the data that have been collected are difficult to interpret, since each population is highly polymorphic; there is no population with a simple genetic profile; the best that can be done is to note that the absence of certain genes makes Europeans and Africans seem relatively close, as against Far Eastern people; this gives a tree that is closer to that based on the Rhesus system than to that based on the Gm system.

Individual Diversity. Population Diversity

At this point, the reader must surely have the impression that the accumulation of increasingly precise data and its treatment by increasingly complex processes have only made it more difficult to classify populations. The clear-cut division presented by the geography textbooks of our childhood into whites, yellows, and blacks is no longer acceptable. Has scientific research gone wrong?

The role of science is not to invariably give clear answers to all questions. Some questions are better not answered; in replying, even partially or imprecisely, to an absurd question, one becomes party to mystification and to an abuse of trust.

If the classification of humans into more or less homogeneous groups, of the kind referred to as "races," had real biological meaning, the role of biology would be to establish this classification in the best possible way; but this classification has no meaning. For it to do so, the history of humanity should have unfolded in the manner illustrated by the tree in figure 6: a series of successive splits. In reality, our present human populations were never separate for long enough to allow a significant genetic differentiation to take place. Individuals moved from one group to another and we saw that even a tiny trickle of migration can have major consequences.

Populations can be compared with respect to specific traits and the differences observed can then be analyzed; in some regions, the degree to which relatively separate populations have differentiated from each other can be studied; but this work does not lead to an objective classification of the human species into "races." The best proof of the uselessness of attempts at defining races was no doubt provided by American researchers R. Lewontin (1974) and M. Nei (1975). They tried to analyze the total global diversity of all humanity into: (1) differences between the major commonly defined groups (whites, yellows, blacks), (2) differences between nations within one of these groups, and, finally, (3) differences between individuals within the same nation: the proportions are respectively 7, 8, and 85 percent. In other words, the total diversity found between humans is reduced by an average of only 15 percent when, instead of humanity as a whole, one considers only those people who belong to the same nation.

This result is worth thinking about: it is not between human groups, but between individuals that the level of diversity is at its highest. Of course, my friend Lampa, a bedik peasant from eastern Senegal, is very black and I am almost white, but some of his blood systems are perhaps closer to mine that are those of my next-door neighbor Mr. Dupont. Depending on which criterion I choose as a basis for comparison, the distance between Lampa and me will be greater or smaller than the distance between Mr. Dupont and me. The result obtained by Lewontin and Nei indicates that the biological distance which separates me from Mr. Dupont is, on average, only a fifth smaller than the distances which separate me from Lampa, or from one of my Japanese or Indian colleagues, or from an Australian desert hunter-gatherer. Does this small difference deserve all the attention we have been giving it for centuries?

FIVE

Evolution and Adaptation

THE UNITY of the living world and the interrelatedness of all species during the slow evolutionary process have come to be generally accepted as facts by all but a few fundamentalist groups who object, not on scientific, but on religious grounds. Where does this human species of ours come from? Before trying to reconstruct its past, consider the kind of universe in which it finds itself.

The Universe. Man

According to astronomers, the universe which is accessible to our observation (and our tools of observation are now infinitely more powerful than our senses) occupies a vast space with a radius greater than ten billion light-years.[1] Within this space, matter is not distributed uniformly; it is aggregated into "lumps," the galaxies, of which there are many billions.

Our own galaxy is none other than the Milky Way; it is a major feature on our sky, not because it is larger than the others but because we see it from the inside, whereas the closest outside galaxy is 200,000 light-years away. The Milky Way, shaped like an im-

---

1. One light-year is equal to the distance covered by light in one year, that is, about ten thousand billion kilometers.

mense disk bulging at the center, comprises about a hundred billion stars.

Our star, the sun, is one of these; apart from being our particular star, it is in no way different from this vast throng; it is situated relatively far away from the center of the galactic disk, about 30,000 light-years, which is two-thirds of the length of the rays of this disk.

Our planet, the earth, is one of the nine large planets that accompany the sun; third in order of distance from the sun, it is only 150 million kilometers away from it.

This entire universe is on the move: the galaxies are going farther away from each other as though they had been projected outward by an initial explosion, perhaps by the famous "Big Bang" that is thought to have created our universe; they are themselves revolving, bringing the stars with them in a dizzying whirl; the corteges of planets take part in this complex set of rotations as they revolve around the stars, with each planet revolving on its own axis and acting as a pivot for its own satellites. Our earth turns on its own axis once a day, around the sun once a year, around the center of the galaxy once every 250 million years. Each of us is turning around the earth's center at a speed of 1,600 kilometers per hour, around the sun at 50,000 kilometers per hour, around the center of the Milky Way at a speed of one million kilometers per hour.

This movement is undoubtedly not "perpetual"; it had a beginning and it will have an end. According to the "Big Bang" theory, the universe as we know it is about 15 billion years old; the age of our solar system can be estimated, with greater precision, to be about 5 billion years.

Five billion times our planet has revolved around the sun; but, at each rotation, some changes occurred: around this infinitesimal agglomerate of matter, an atmosphere was gradually formed by gases from volcanic eruptions; the oceans resulted from condensation of the steam; thanks to the energy of the ultraviolet rays of the sun, simple molecules associated, became increasingly complex and acquired new properties until the appearance, about 3.5 billion years ago, of molecules possessing the strange and awesome power of making other molecules and of reproducing themselves: "life" was beginning.

There is no fundamental difference between the "living world" and the inanimate world; they are both made of the same matter and are subject to the same forces and constraints. What we call "life" came about, not brutally through some resounding miracle, but slowly, gradually, laboriously, and hesitantly through the very dynamic of inanimate matter.

It was long believed that what distinguished the "living world" was its ability to defy or transcend the "second principle of thermodynamics." According to this principle, stated by Carnot at the beginning of the nineteenth century, all forms of energy must, of necessity, deteriorate; although strictly accurate in the case of a finite system, it can, through a dangerous extrapolation, be used to predict a general waning of the strength of the universe and its ultimate collapse into unstructured grayness; physicists characterize this process of decline and general degradation by saying that "entropy" is constantly increasing. Living matter appears on the contrary to be capable of maintaining its structure and even of evolving toward ever greater complexity and efficiency; its "entropy" may be decreasing. However, over the past ten years, this opposition has come to seem less clear-cut. Recent research in thermodynamics, and the work of I. Prigogine (1977) in particular, has shown that Carnot's second principle gives but a simplistic view of a multifaceted property of matter: as soon as material systems are sufficiently complex, they spontaneously structure themselves in such a way as to minimize the production of entropy, which is exactly what living matter does; instead of a division along the lines of "living" and "non-living," one should think in terms of continuity from the more to the less complex; thus, the unity of the whole tends to be reestablished.

This rediscovery of unity does not mean that our picture of life is somehow restricted; it merely means that we have become more aware of the complexity of the "laws" governing matter. Of course, biologists are always trying to describe the functioning of living organisms in terms of concepts that have been elaborated by physicists, but, in return, physicists look to biologists for a better understanding of some of the more surprising aspects of matter.

This deeper unity in no way negates the stupendous diversity of the products of this exuberant world, which can be described either as "living" or as "hyper-complex."

The number of species to which this earth is host has been estimated at one and a half million; their diversity of appearance and function gives an impression of heterogeneity; what do a seaweed and a sea gull, a jellyfish and a human being, have in common? The evidence for the relatedness of species becomes overwhelming when one looks, not at external appearances, but at basic internal structures. The processes by which all organisms insure their development and survival, both individually and collectively, are extremely similar: the cells of all of them transfer energy by means of the same chemical compounds, especially adenosine triphosphate; the cell membranes all have the same structure; energy storage is accomplished by the same products, fat or carbohydrates, and the necessary reactions are catalyzed by proteins of very similar structure; above all, the production of the various proteins based on the information in the genetic code is accomplished through a mechanism determined by a universal "code" that is applicable to all. It seems highly improbable that these traits could be present in all living organisms if these do not have a common origin. With near absolute certainty, we can declare that the living world is one.

Darwinian Evolution

In three billion years, the capacity for differentiation characteristic of living things has led from the first whisperings of life, in the form of a few molecules endowed with the power of self-reproduction, to a proliferation of organisms with multiple powers, all of which are wonderful and some disquieting: thus, in man, the power of becoming aware of his own gifts and of developing them to the point where he has the power of destroying life itself. The long journey traveled can be conjectured and, thanks to the progress of knowledge, it can now be chronicled with increasing precision; however, description is not enough, we need to understand; we need to imagine the process which led to this chronicle.

Transformationist theories were outlined as early as the eighteenth century by Maupertuis or Buffon and further developed at the beginning of the nineteenth century by Lamarck. However, it was not until after the publication of Charles Darwin's *Origin of the Species* in 1859 that these theories were widely debated. The idea that species change and evolve into new ones did not originate with Darwin; other researchers had proposed this thesis before him, but it had not reached the general public. Darwin was original in that he proposed the mechanism of "natural selection" as an explanation for evolution; Darwinism is therefore not to be confused, as it still is by many, with transformationism. Darwinism is the explanation of the transformation of species in terms of a "struggle for existence" that eliminates the least fit and preserves the "best." The essence of this theory depends on two observations:

1. Animal breeders succeed in changing animal species; to do so, they select animals for reproduction based on criteria that are often chosen for economic reasons (e.g., milk-yield in cows) or esthetic ones (e.g., type of coat in dogs). "Artificial selection" has proved to be very efficient; virtually any trait can be modified in this way; in some cases, the transformation is relatively rapid.

2. In almost all populations, more individuals are born than can possibly survive, since resources are limited; those that do reach reproductive age have undergone a process of "natural selection" that eliminated the weaker ones.

The first observation proves that the traits affected by artificial selection are transmitted from parents to offspring; in fact, those differences between traits that determined the breeders' choice are reproduced, at least partially, in the offspring: a cow with an exceptionally high milk-yield has progeny with a milk-yield that is, on the whole, higher than average.

The second observation indicates that certain traits (e.g., physical strength, resistance to the environment) affect an individual's capacity for survival and procreation and condition its "selective value"; these traits are selected naturally, they must therefore spread progressively throughout the population; from generation to generation, this population thus becomes transformed; in other words, it evolves.

The very choice of the term "natural selection" shows that Darwin meant to stress a fact that seemed fundamental to him: evolution uses the same material as animal breeders and this material is none other than the differences between individuals. It is to the extent that a trait is variable that one can hope to modify it through selective breeding; it is to this extent also that it will evolve spontaneously over generations.

In the natural process, the deliberate intervention of the breeder who chooses the animals for reproduction is replaced by the competition between individuals for the resources necessary for survival and for reproduction.

Presented in this manner, the Darwinian theory seems obvious. However, it raises certain problems:

1. The variability of individuals, without which evolution could not occur, is systematically reduced by the very process of natural selection. This favors those individuals who are closest to a certain ideal type, which in turn is determined by the conditions imposed by the environment. Little by little, their descendants become closer to the type. The population gradually becomes homogeneous, thereby depriving natural selection of its raw material. Furthermore, the gemmule theory, which Darwin accepted as an explanation for the transmission of traits from parents to children, implies a reduction in this variability with each generation since the offspring represent the average between the two parents. Observation shows us that, quite to the contrary, the variability of most traits is extremely high. How is it maintained? Above all, how did it originate?

2. When one refers to the selection of the "fittest," the word "fit" has to be understood in a very precise and rather limited way: it refers to the ability to survive and procreate. A trait is selected only to the extent that it directly influences this type of fitness. The expression "selective value," which we introduced earlier, must

not be misinterpreted; in particular, the word "value," with all its connotations, is likely to be misleading: we can say that the best individuals win the battle for selection provided that the best are defined as those who are best equipped for winning. At that point, the Darwinian theory is no longer just obvious, it is tautological.

3. The competition or struggle for survival is not merely between individuals, but also between whole populations belonging to the same as well as to different species, living in the same environment. A trait that handicaps an individual in his struggle with his fellows may benefit a population in competition with other populations. This is the case, for example, with genes for "altruism" which seem frequent in certain animal populations. These genes handicap their carriers because they lead them to sacrifice themselves for the common good; they benefit the community because they provide it with unselfish defenders. Depending on the intensity of competition at these two levels, such genes will multiply or disappear, and it will be impossible to explain their destinies in terms of their individual selective values.

4. However, the most serious weakness of the Darwinian theory, as its author knew perfectly well, is the lack of a correct explanation for the transmission of traits. In trying to understand the evolutionary mechanism of a group, Darwin was hampered by his inability to explain the elementary process, the making of an individual. He resorted to the gemmule theory in 1868, but only as a "provisional hypothesis," in order to fill the gap. We remember that, three years earlier, in 1865, Mendel had imagined the genetic model which we now know to be an accurate representation of reality. His ideas, however, had been given little attention in spite of the fact that he had communicated them to certain scientists; they were too innovative to be accepted just then. Darwin, since he did not know them, was obliged, like his predecessor Lamarck, to assume the heredity of acquired traits and to base his theory on a false hypothesis, as can be clearly seen from this sentence describing natural selection: "Can it be doubted that individuals possessing some advantage over others would have a better chance of surviving and of procreating their own type?"

We now know, as we insisted in chapter 1, that the mechanism of sexual reproduction does not allow an individual to "procreate his own type." He can transmit to his offspring only half of the collection of genes that determined his "own type," which is an entirely different mechanism.

When, in 1900, Mendel's discovery was finally understood and accepted, biologists were unable to see that it filled a gap in Darwin's theory. On the contrary, the two theories seemed to be in radical contradiction with each other; the contradiction seemed so complete that unusually violent polemics developed, especially in Great Britain. Closer analysis gradually showed that this conflict hid an underlying complementarity. Mendelism did not destroy Darwin's theory; instead, it made it possible to give it a solid basis and to develop it. Most of those responsible for this reconciliation were mathematicians who developed a remarkably coherent synthesis: neo-Darwinism.

Neo-Darwinism: A Convincing Synthesis

At the beginning of this century, the split between evolutionism, as it had been developed by Darwin's successors, and genetics, as it had begun to emerge since the rediscovery of Mendel's work, was total. The evidence for the progressive change of populations and individuals, for their adaptation, seemed to be in conflict with the stability of the genes which constitute the indivisible and almost inalterable "atoms" of the genetic inheritance. Those who accepted the Mendelian model appeared to be anti-Darwinians and therefore dangerous iconoclasts. It was many years before it was discovered that the observations on which Darwinism is based and the explanatory models developed by geneticists could be integrated into a coherent synthesis. At first, this synthesis, neo-Darwinism, which relied heavily on mathematics, was not well received (until 1937, one of the major English biological journals, *Biometrika*, refused all Mendelian articles—Sentis 1970); gradually it, in its turn, acquired the status of an official theory, the least criticism of which seemed heretical.

We saw that, as early as 1908, a mathematician and a biologist brought to light a rather remarkable aspect of the behavior of genetic structures, which came to be known as the "Hardy-Weinberg rule." However, many other theoretical results can now be worked out, due to the fact that certain concepts have been precisely defined.

The essential difficulty, in this domain where the objects under study cannot usually be observed directly, is, in fact, with concepts. There can be no serious argument or valid result unless the words used have first been rigorously defined. This difficulty is encountered in the case of the central concept, that of evolution. What do we mean when we say that a population evolves? The individuals who make up this population cannot "evolve." Except in the case of extremely rare mutations, the genetic heritage of each individual remains the same throughout his existence: from conception to death this heritage is fixed. There is no evolution from father and mother to son and daughter, because the latter are entirely new creations; we insisted on this point in chapter 1. Because of the way in which sexual reproduction functions, the absolute uniqueness of each new being is routine necessity and not the extraordinary miracle it once seemed to be.

Finally, that which evolves is neither the individual nor the collection of individuals in a population but the complement of genes which they carry. From one generation to the next, this complement changes because of many factors:

1. Mutations produce new genes. Though they occur very rarely, they are, for a species taken as a whole, the only source of real innovation. A mutation sometimes concerns one point only, that is, it may affect only one elementary trait, a single gene; however, it can also bring about the restructuring of an entire chromosomal zone, in cases where the breaking of certain chromosomes leads to their reassembly in a different order. Even the number of chromosomes can be modified when, for instance, two of them fuse into one. When any of these accidents occur in the sex cells of an individual, the genetic heritage transmitted to his descendants is enriched by a new trait. The variability of the species as a whole is increased by it.

2. A new gene can be introduced into a population by an im-

migrant from another population within the same species. These migrations, which are especially frequent in the human species, play a major role in the maintenance of variability within each population.

3. The genes thus introduced, whether by mutation or migration, can have either a beneficial or a harmful influence on the capacity of the individuals who carry them for survival and procreation. This influence depends, of course, on the "environment," that is, on the conditions in which the group lives, as well as on the other genes carried by that individual.

4. As we saw in chapter 2, the finite size of the group causes a random variation in gene frequency; the smaller the group, the greater the role played by chance; this phenomenon has been called "genetic drift."

5. Finally, mating patterns of individuals within a population can influence the process of gene transmission: for instance, if people with an a gene never marry people with a b gene, the ab heterozygotes disappear.

The aim of neo-Darwinism is to review these various factors, to define their influence on the destiny of a gene and to set out clearly the rhythm at which the transformation of genetic structures takes place. Naturally, a process of such complexity can be studied only after it has been represented by models with some degree of realism but which must be sufficiently simple to be treated mathematically.

The simplest approach involves the study of an elementary trait in isolation from all other traits; suppose the Rhesus system is studied in this way. We saw in chapter 4 that this system, for which only two genes were indentified initially, corresponding to the positive and negative traits, actually depends on about twenty genes, which we will label $R_1 \ldots R_{20}$. A population in a particular generation is characterized by the frequencies $R_1 \ldots R_{20}$ for these genes; its evolution between this generation and the next is defined by the changes in these frequencies, which become $p'_1 \ldots p'_{20}$, due to the effects of the various factors that we mentioned. The frequencies p_1, for instance, will decrease if several individuals carrying the gene R_1 die before procreating; this event may, of course, occur independently of the presence of the R_1 gene in their genetic heritage; it may be caused by an accident or by genes affecting other traits. However, in a sufficiently large population, we may assume that the R_1 gene will be associated sometimes with beneficial genes and sometimes with harmful ones and that those influences that are external to the trait will balance each other out. If we define the selective value of individuals as their capacity for transmitting their biological heritage, we can then calculate the average selective value of all carriers of the R_1 gene, regardless of the associated genes, and thus define the selective value of this gene. When this value is lower than the average, the frequency P_1 decreases in each generation; when it is higher, this frequency increases. We will not go into the details of the mathematical arguments that permit the calculation of the relationship between the selective values, thus defined, of genes and changes in frequency (Jacquard 1974). The results of these analyses make it possible to give a quantitative description of the evolution of a group and to calculate the rhythm at which a new gene, R_{20} for instance, introduced through mutation or immigration, may eventually spread, gradually replace the other genes, and, when the frequency p_{20} reaches 1, become the only gene for the Rhesus system in the population.

Let us consider in detail one general theorem which is important, not so much because of its own formulation, but because of the consequences that can be derived from it, and which came to be known as social Darwinism. At one time these ideas gave renewed life to a certain vision of the relationship between nature and man, and between humans.

The theorem itself was developed in 1930 by Sir Ronald Fisher, founder of neo-Darwinism with fellow-Englishman J. B. S. Haldane and American Sewall Wright. Fisher (1930) gave it the rather pompous title of the "fundamental theorem of natural selection"; this wording shows how convinced he was that Darwinism had finally been given the mathematical formulation without which the sciences do not seem to be "exact."

This theorem states that "the increase in the average selective value of a population is proportional to the variance of the selective values of the genes which constitute the heritage of this population." We need to get past this rather obscure formulation to reach the meaning it attempts to convey.

(First, note that the "variance" of a group of measurements is defined as the average of the squares of the differences between these measures and their average; it tells us about the dispersion of the whole; a positive number by definition, it becomes greater according as the measures are more dispersed.)

The biological objective of any group of living things is to struggle against the erosion of the passage of time, to survive and to adapt itself in the best possible way to the conditions of the outside world. A group's adaptability must be measured in terms of the selective value of the individuals belonging to it: the higher this value, the greater the group's adaptability. The average selective value changes due to the influence of various factors on gene frequency. Fisher's theorem shows us that this influence leads of necessity to an increase in the average selective value and that this increase occurs all the more rapidly the greater the variability of individuals within the group.

Everything therefore seems to be for the best: natural selection cannot but improve the situation. The common good is best served by allowing this type of selection to operate freely. It is easy to see how this theorem came to be used as a justification for all sorts of wild statements on the excellence of the natural order of things.

An Unwarranted Extrapolation: Social Darwinism

The extraordinary impact of Darwin's theories certainly cannot be ascribed solely to the quality of his scientific thinking: the first edition of *Origin of the Species* was sold out on the very first day and a much larger second edition had to be rushed to press. A society embraces a new theory with such enthusiasm only when this theory helps, even if unwittingly, to solve some of its problems (Achard et al.). Remember how certain theologians used the discovery of "homunculae" in the sperm as a justification for the concept of original sin.

In the industrialized England of the second half of the nineteenth century, fortunes were being built up on the profits made from mines and factories where the workers were paid barely enough to survive. In order to limit costs, certain children working in the mines were brought up to see the light of day only once a week. In this same society, however, religion was very important and was the supreme point of reference. A certain malaise could scarcely be avoided. England, along with other European nations, was involved in what was for them the exhilarating adventure of colonization. This adventure led to the annexation of entire countries whose inhabitants were considered inferior to those of the conquering white powers. The victory of the latter seemed assured and irreversible. Their domineering attitude did not fit in very well with the tenets of their official religion preaching brotherly love.

Along came a scientist saying that the progress of the living world is a result of the "struggle for survival." The improvement of each species, the evolution from one species to another more advanced one, can be achieved only through the elimination of the least fit and through the victory of those who have received a better heritage. This rule is not of human invention, it is a dictate of Nature. It is only by adhering to it that the common good can be fostered. This scientific statement is, of course, based on the observation of animals; it concerns only those biological traits that affect survival and the capacity for procreation. However, it was immediately understood to be a justification (and the most objective possible, that of Nature itself) for competitive behavior: since struggle is necessary for the progress of the human species, no matter how brutal this struggle, its outcome must be considered beneficial. The concept of the "will of God," which had permeated the Middle Ages, was replaced by that of the "will of natural selection." If whites defeat blacks, it is because they are better; it is normal, and better for the human species, that the former should oust the latter.

It would be unfair to attribute these extensions of his theory to Charles Darwin himself. However, it does seem that he largely succumbed to the temptation to extend the process of "struggle for life" to the social domain. Pierre Thuillier (1974) has gathered a certain number of quotations which show the extent to which Darwin was tempted by eugenics, while at the same time dreading its excesses: "Man must continue to be subjected to tough competition"; it is necessary that "all laws and customs that prevent the success of the most capable be abolished."

It is important to realize fully the influence exerted by such arguments on public opinion even today. Racism can still develop in the name of a supposed scientific truth; it is in the name of this "truth" that people sometimes attempt to justify the most shocking inequalities in treatment. It would be too easy to ridicule these pretensions to scientific justification by quoting some of the many wild statements made by people who have referred to science without knowing anything about it. We will concentrate instead on a few sentences written by an eminent biologist, a Nobel prizewinner in physiology and medicine:

> In the interest of race preservation, morally inferior beings should be carefully eliminated with an even greater degree of strictness than at present . . . We should—and it is our right—put our trust in the best among us and commission them to make the selection that will determine either the prosperity or the destruction of our race. (Thuillier 1974)

That these lines were written by Konrad Lorenz in the Germany of 1940, at a time when the concentration camps were already functioning, makes matters even worse. Disregarding preestablished morals, let us try to judge only the logical content of these propositions. Is there any real basis for a "social Darwinism"?

Before replying, we must first establish the exact meaning of the words. Those used by Lorenz are revealing: "inferior beings," "the best among us." His entire argument is based on the assumption of a hierarchy within populations. A hierarchy can of course be created within a population; to do so, all one needs is to choose arbitrarily certain quantitative or qualitative traits (height, intelligence quotient, skin color, annual income, etc.) and a formula that makes it possible to synthesize these various criteria into a single measure: those individuals who score the highest marks are, by definition, the "best." However, this classification is relevant to the "improvement" of the group only if offspring show a tendency to score like their parents, in other words, if this score is a heritable trait. The difficulties surrounding this concept of "heritability" will be dealt with in the next chapter; for the moment, we merely point out the following rather obvious fact: the elimination of "inferior beings" is of long-term interest only if their children are also likely to be inferior and if this likelihood is due to biological and not social factors. However, we do not have the slightest proof of this.

Above all, this argument is quite different from Darwin's. His fundamental criterion for judging success in the struggle for existence was the number of children produced. The only correct definition for selective value, which we saw to be central to neo-Darwinism, is based on the number of genes transmitted by each individual to the next generation. This concept has a clear biological meaning but it cannot, without great caution, be applied to the analysis of the evolution of our own species: Leonardo de Vinci, Beethoven, or Lenin, all of whom seem to have had no children, had selective values of zero; in the Darwinian sense of the word, they were "inferior beings" (Lerner 1968:33).

In spite of its name therefore, social Darwinism is by no means an extension of Darwin's statements on natural selection, which regulates the evolution of the living world. It implies an entirely different approach which favors a deliberate and voluntaristic choice in favor of artificial selection. It is only by blantant misuse of language (even if Darwin himself fell into the trap) that the acceptance of a social or political order implying inequality, oppression, and exploitation could be presented as a consequence of natural processes. If one wishes, without hypocrisy, to found a social Darwinism, one must draw on those concepts and arguments that permit the elaboration of a technique for species improvement; it is no longer a question of the natural order of things, but of man's potential for changing it; we will see in the next chapter the difficulties encountered on this path.

A Radical Critique: Non-Darwinism

We mentioned some of the objections which Darwin's initial theory encountered; the developments added by neo-Darwinism to this theory made it possible to counter most of these objections; the source of variability was identified: mutations bring new genes to each generation; the mechanism for the transmission of traits was elucidated: Mendel's model, and later the discoveries made by cytogeneticists showed how the chromosomes, the carriers of the hereditary material, duplicate. However, the classical neo-Darwinian theory offers no satisfactory explanation for the long-term maintenance of a high degree of polymorphism: mutations are so infrequent that deleterious genes must be very rare. The simplest model for explaining the simultaneous maintenance of several categories of genes for the same trait is that which assumes the selective advantage of heterozygotes: such seems to be the case with sickle cell anemia, as was explained in chapter 2; the loss of S genes due to the death of homozygous SS children is compensated for by the loss of "normal" genes due to the death of individuals who, not being carriers of the S gene, have a weaker resistance to malaria. A "polymorphic" balance is therefore possible, but only because of the very high mortality among children. For this single trait, polymorphism is maintained, in some regions of Africa, through the elimination of 10 percent of the children (one quarter of which die of sickle cell anemia and three quarters of malaria). Such a heavy genetic burden could obviously not be tolerated for numerous traits simultaneously.

About fifteen years ago, it was generally believed that populations were homogeneous for most traits. A relatively small proportion of these traits could be kept polymorphic through simple selective mechanisms (such as the advantage of heterozygotes) or more complex ones (such as that which is based on the variability of selective values as a function of gene frequences, a phenomenon which Claudine Petit, (1976), among others, described in *Drosophila*).

The systematic use of electrophoretic techniques (that is, techniques for differentiating between molecules as a function of their migration rates in an electric field) has shown that biological reality is definitely not in accordance with this vision. One of the advantages of this technique is that the sample of proteins that it allows us to study can be considered to be reasonably representative of all the proteins produced by the organism (while this degree of representativity was by no means assured by the blood groups which, up to then, had been our sole source of knowledge of human polymorphism). Furthermore, this method is inexpensive and the equipment required is simple: the protein is placed on a gel and subjected for several hours to an electric field; under the influence of this field, it moves; the distance traveled depends essentially on its electric charge; specific stains make it possible to detect its final position on the gel in the form of a dark band.

Using this technique, samples from several hundred individuals can be studied: the same protein, with the same role in all the organisms studied, with no detectable difference in function, may very well give rise to bands whose positioning varies according to the individual. Thus, their functional uniformity hides a heterogeneity which signals the presence of different genes. Mutations that could not previously be detected can be revealed by this technique.

When a sufficiently large number of subjects are studied, a few individual variants are usually found. In order to define the concept of polymorphism, no account is taken of exceptional cases which have scarcely any significance for the population. By a generally accepted convention, it is assumed that a trait is "polymorphic" when at least 2 percent of the subjects are heterozygotes.

Research has shown that this polymorphism is much more common than predicted in both animals and humans: to characterize it, let us say that, in most populations, at least 40 to 50 percent of traits are polymorphic, or—looking at the same phenomenon from a different angle—that at least 15 percent of any individual's traits are heterozygous (Lerner).

This finding goes contrary to many widely accepted ideas. Since it had not been foreseen by previous theoretical developments, the theory now has to be revised. Two avenues can be explored:

1. The first attitude involves shelving, at least provisionally, the concept of the selective value of the various genes. When a gene is produced by a mutation it may, as is frequently the case, be very harmful to its carrier: modifying, for example, the structure of an enzyme, it renders it inoperative, seriously perturbs metabolism, and makes its carrier incapable of surviving and reproducing. Genes

of this kind are eliminated as soon as they are introduced to the collective heritage. Therefore, they do not contribute to polymorphism. The other mutants, which are compatible with the normal functioning of the organism, no doubt exert some influence, whether favorable or not, on it. However, this influence is not something that can be measured. It is considered good scientific practice to consider them to be "neutral" and to study their evolution while assuming their neutrality as a working hypothesis. This attitude has been adopted, especially for the past ten years or so, by geneticists such as Motoo Kimura (1971) and Masatoshi Nei (1966) among many others, who have made extensive use of mathematics in their research work. Some have provocatively maintained that this orientation implies a non-Darwinian theory of evolution. Such an extreme formulation is in danger of distorting the real intention behind this research. In reality, what they are doing is emphasizing the role of chance, the evolutionary factor introduced by Mendel's discovery, and making as little use as possible of the imprecise Darwinian concept of "selective value." It so happens that the explanatory power of those models, which assume equivalence between genes whose destiny is no longer considered to be a function of qualities inherent to them but rather of contingent and unforeseeable events, has proved to be quite remarkable. A large proportion of the findings, in particular concerning the distribution of gene frequencies, are compatible with the inferences that can be drawn from this hypothesis. Our vision of the evolutionary process is profoundly changed by it. Its rhythm can no longer be seen as being dictated by the intensity of selective pressures, but by the frequency of mutations. Chance, not necessity, therefore plays the primary role.

2. The second and no doubt more "classical" option involves making the initial theory less simplistic so as to allow it to more adequately reflect observed reality. Here, the researcher has a wide choice because the first models of neo-Darwinism were consciously simplified. The most important research in this line has involved the consideration of several elementary traits simultaneously. Some entirely unexpected properties were discovered by American teams such as those of Karlin (1975) and Feldman (1977)

at Stanford University or of Lewontin (1974) at Harvard. When
two or more traits simultaneously affect the selective value of individuals, the behavior of the genetic structures of a population
can be, to say the least, paradoxical in some respects. Fisher's fundamental theorem is usually not verified: the process of selection
does not necessarily bring about an increase in the average selective value. The number of stable equilibria that can be shown to
exist, whether through calculation or through computer simula-
tions, increases very rapidly with the number of traits considered;
when the conditions under which a population lives remain constant for many generations, its genic structure tends toward one of
these equilibria, but, since these are extremely numerous, the outcome depends as much on the initial genic structure as on the selective pressures of the environment (in the same way that a mar-
ble dropped on a bumpy surface ends up in one of the hollows;
depending on its starting point, its arrival point can vary greatly).

It is therefore very difficult to explain the evolutionary path of
a trait as a function of the links between this trait and the selective
value of individuals. A new parameter is introduced, the location
of the genes on the chromosomes; the degree of evolutionary in-
terdependence between two traits increases as the chromosomal
distance between the corresponding genes decreases.

Consider once again the famous example of the giraffe's neck:

— For Lamarck, giraffes, in their need to reach up for food, stretch
their necks; this acquired trait is transmitted to their descendants
and, gradually, the species acquires a longer neck;
— for Darwin, this process is accelerated by the "struggle for survival," which favors giraffes who have either received or acquired
necks that are longer than average;
— for neo-Darwinians, acquired traits cannot be transmitted;
therefore the differences between the lengths of necks in a population of giraffes correspond to differences in genetic complements;
the fact that longer-necked giraffes contribute more to the process
of gene transmission gradually increases the frequency of the genes
responsible for increased length;
— to many present-day geneticists, it seems difficult to isolate this
one trait; selective pressures have affected numerous traits simul-

taneously; for long-neck genes to become widespread, it was sufficient that they be associated by chance with genes favoring metab-
olisms for entirely different organs; the long-neck genes themselves
were not necessarily beneficial.

A human trait, which we discussed in connection with the definition of races, seems to have been subjected to indirect selective
pressures of this kind. Populations with very dark skin, very rich
in melanin, are located in regions where the climate is exceptionally hot: Melanesia, the Indian peninsula, Central Africa. On account of this very color, the regulation of body temperature is more
difficult: a black-skinned person absorbs 30 percent more solar energy than a light-skinned one. Dark skin would be more beneficial
to a person living in Sweden than to someone in Senegal; the former could use the extra calories to fight the cold while the latter
has to eliminate them to fight against the excess heat. It is true, of
course, that light has other effects besides supplying energy: burns
and cancer induction, against which melanin provides some degree
of protection, antirachitic vitamin D synthesis which melanin, depending on the circumstances, may either help or hinder. At any
rate, these effects are not intense and it would be difficult for them
to exert a significant selective pressure; moreover, they are linked
not to the overall intensity of solar radiation, but to the intensity
of rays with a very short wavelength and the distribution of these
rays is only weakly correlated with the distribution of skin colors
(H. Blum).

It therefore seems impossible to assume that skin color is a
product of a simple adaptive process. The location of dark-skinned
people in certain regions can hardly be attributed to the chance
effects of migration; it would be too much of a coincidence. Perhaps complex selective pressures, involving other seemingly unrelated traits, and which we cannot at present envisage analyzing,
were at work.

In conclusion, note that global models for the action of natural
selection, which take account of the interaction of numerous genes,
actually show the evolution of a trait to be dependent on phenomena with which it is in no way connected. A particular gene may

spread throughout the population, another one may be eliminated, not because of some beneficial or harmful property of its own, but because of its chance association with genes governing entirely different traits. This description amounts to saying that evolution is subject to "chance." This much-used word is likely to cause confusion; remember that we defined chance, not as the absence of causes, but as the absence of identifiable causes or, to put it differently, as the convergence of "independent series of causes," in the words of Augustin Cournot. (When we say that roulette is a game of chance, we do not deny the existence of the many deterministic forces governing the movement of the ball, but we do admit our inability to analyze their effects.)

The two research orientations that we mentioned, one assuming the neutrality of the various genes and thus dispensing with the concept of selective value, the other taking the complexity of reality into account, linking selective value to the genotype as a whole, amount to the same thing in the last analysis: whether chance is introduced as an explanatory factor, or whether it results from the complexity of the determinism involved, it is on it that we finally rely to describe evolution.

The Illusion of Type: The Reality of Dispersion

Much of our reflection on the living world is based on our belief in the existence of a "type"; this old Platonic concept allows us to classify and judge things, animals, and people. Before speaking of "Persian cats," "German shepherds," or "French people," we have to specify what a cat, a dog, or a person must be in order to belong to these categories; in other words, we have to define a type; thereafter, each individual can be judged according to how closely he resembles this type. This attitude of mind was further reinforced by the way that Darwin's thinking was interpreted for a long time: in each environment, the "best" win and the population gradually adapts; this adaptation seems perfect to us when, for each trait, all individuals carry those genes that were retained by natural selection. This way of seeing things is very deeply ingrained in us: for a given species, in a given environment, we tend to think that there is an optimal biological or genetic response, which nature is capable of gradually specifying and achieving.

Population genetics shows that this is only an illusion. Nature does not embody a process leading to the convergence of individuals toward an ideal type. It seems to have an entirely different strategy: that of preserving diversity. Doubtless, in the short term, certain genotypes have a higher selective value; their presence in greater frequency is beneficial to the group. However, the group's capacity for evolution depends on the diversity of these genotypes: the present depends on the average, but future developments depend on variance.

The living world that we observe is not the culmination of a series of deterministic events, which could not fail to lead to the state that it has now reached; it was not inevitable. At each instant, reality is pregnant with an infinity of possibilities. The "laws of matter" or "the laws of evolution" intervene to make each of these possibilities more or less probable but they do not dictate the result of the lottery; at least, the "laws" that we can identify are not capable of exerting a constraint of this kind. From this infinity of possibilities, a single reality will emerge, the choice of which must be attributed solely to "chance"; and this reality is not necessarily one of the possibilities with the highest probability. All the details of the phylogenetic tree were not predestined right from the origin of life; and the new branches that may still emerge from it cannot be predicted.

SIX

Species Improvement: What Improvement?

"**Y**OU CLAIM that it is impossible to improve the human species. Nonetheless, Man has succeeded in improving numerous animal or plant species. You can hardly question, for instance, the improvement in breeds of horses."

"If you were a horse, would you really consider it an improvement?"

Though somewhat extreme, this recent dialogue with a journalist seems to me to present the problem quite well. The successes of artificial selection are undeniable. Over the centuries, we have, through deliberate intervention, transformed certain species. However, before attempting to change our own species, we must specify:

— the objectives to be pursued,
— the techniques to be used to attain them.

This chapter will deal with the question of techniques.

Success

The founding of agriculture was quickly followed by the domestication of animals; that of the dog, the hunter's auxiliary, may even have preceded it, since it is believed to have begun more than twelve thousand years ago; that of horses, cows, and sheep occurred later, probably four to six thousand years ago. This domestication was gradually accompanied by an attempt to enhance those qualities that appealed to breeders; it seems that attempts at preserving traits pleasing to breeders predated attempts at increasing useful traits, for instance, milk- and meat-yields.

Not until the eighteenth century was a systematic attempt made at improving certain useful traits in livestock through the use of controlled crosses. This attempt, though largely successful, had some indirect and harmful consequences, especially in making the animals less robust and thus, like the labour of Sisyphus, it needs to be constantly begun again from scratch.

From the thirties onward, with the progress of population genetics, the methods developed empirically by breeders began to get a solid theoretical basis. Fisher's initial analyses, to which we have already referred, were followed by a proliferation of models which provided clearer guidelines for the breeder's choice.

One cannot but admire the results. All the animal traits on which our food supplies depend have been improved, sometimes to a spectacular degree: milk-yield from cows, growth rate of pigs, and egg production by hens have all increased dramatically; in "traditional" countries, a cow yields 400 kg per year while in the United States the average yield reached 4,275 kg in 1955 and had risen to 5,500 kg by 1967.

The increase in cereal yield has been even more astounding. The expression "green revolution" has been used to describe the transformations which, over the past twenty years, have resulted in an increase in food supply that is almost proportional to the explosive increase in the human population (not without creating numerous problems).

The most remarkable results were no doubt achieved with wheat, in experiments carried out at the Chapingo International Center for Improvement, the "CIMMYT," in Mexico (Genevois). Wheat had been cultivated in virtually the same way for centuries in that country when, after the last world war, this research institute was set up near the National School of Agriculture, 50 kilometers from Mexico City. The average yield at the time was barely 9 quintals per hectare; the annual harvest of 3 million quintals supplied less than half of the country's requirement. The Center's director, Norman Borlaug, selected from the 5,000 varieties cultivated throughout the country those that had the best resistance to wheat-stem rust; he crossed them with a short-stemmed Japanese variety, did several tens of thousands of hybridization experiments, and finally obtained new varieties with all the qualities desired by producers: a plant short enough to prevent lodging, able to endure drought, tolerating large quantities of nitrogenous fertilizer, and using this nutrient to produce heavier and more abundant grain. In ideal conditions, yields of 75 quintals per hectare could be obtained. By 1965, almost all Mexican farmers were using seeds developed by the institute; the total harvest exceeded 22 million quintals.

This success led to an increase in the resources and responsibilities of the Chapingo International Center; gradually, it became involved in the organization of a whole network of centers located in every continent. Thanks to fruitful collaboration and to exchanges of information and seeds, many countries were able to benefit from the achievements of the various research teams: certain varieties capable of yielding 20 quintals per hectare under almost desert conditions were developed. The progress is not only quantitative, it is often qualitative: using ultraviolet rays, Indian researchers induced mutations that ultimately led to new strains of wheat richer in proteins, especially those containing a greater quantity of the amino acid lysine, and they thus increased the nutritional value of wheat.

Similar research on other plants has been equally successful: the various strains of rice grown at the research station of the Pundjab's Agricultural University were yielding an average of 1 ton per hectare in 1965; the development and widespread use of semi-dwarf varieties increased that yield to 1.8 tons in 1970 and to 2.6 tons in 1975. The extraordinary improvement in corn cultivation is well-known; the hybrid varieties developed in the Wisconsin and Iowa experimental stations not only have yields that would once have seemed incredible, more than 50 quintals per hectare, but they are also so uniform that the harvesting can be mechanized.

In light of such successes, one cannot help feeling a sense of triumph, even if the progress achieved has not yet eradicated malnutrition and the threat of famine from large areas of the earth. However, this success is not due uniquely to the efforts of genetic research units set up for the purpose of species improvement; simultaneously, the use of fertilizers was becoming more widespread and farming practices were improving. It is difficult to dissociate the effects of these various causes. Furthermore, improvements to one trait are often accompanied by unplanned and harmful changes in other traits; these "correlated responses" can, in some cases, lead to real catastrophes which threaten to negate the results obtained: D. Hartl (1977) mentions the case of poultry selected over 12 generations for increased shank length; this result was achieved, but at the same time the proportion of hatchable eggs was reduced by a half. Similarly, the selection of cows for milk with a higher fat content causes a reduction in milk production.

Selection, whether artificial or natural, affects the organism in its entirety and not just some traits. The results obtained from selection for certain traits are accompanied by secondary effects which, in the long run, may be far more important than the changes consciously achieved.

Finally, even when the outcome is mostly desirable, there is no proof that the theoretical models which made it possible are an accurate reflection of reality. Scientists often point out the fallacy of "arguments based on authority." They would do well to also object to the "efficiency argument," which is so often used to sort "good" theories from "bad" ones. The success enjoyed by researchers in the improvement of plants and animals is sometimes considered proof that they are the real masters of the "living matter" they manipulate, that they understand the underlying mechanisms, and that their techniques can therefore be efficiently transferred to the improvement of the human species itself. In fact, this success must not blind us to the inadequacy of the concepts used, an inadequacy that the researchers themselves are the first to recognize and deplore. It is now time for us to look more closely at breeders' selective procedures which, after being empirically and laboriously established over the centuries, have been accelerated over the past decades by the discoveries of genetics, which have provided both the prestige of science and the efficiency of theoretical models. What exactly do they consist of?

"Heritability": A Central Concept

First of all, it is important to understand the observations and arguments that underlie the progress of selective breeders and geneticists. The same terms that are called upon to describe the methods successfully used by agriculturalists to ease worldwide hunger are evoked to speculate on the possibility of improving our own species or developing programs for the creation of a "better" human being. The dangers of misinterpretation are considerable.

It is strange to see how carelessly scientists use some of their most essential tools: words. Researchers, who would never use a soiled pipette or a cracked test tube, sometimes feel no scruples about using outdated words that are worn out from having been too often spoken and written, and that have lost all precise meaning from having been used as labels for many different concepts.

The two qualifiers "genetic" and "heritable" are often used as though they were equivalent. A trait that is genetically determined naturally seems to be heritable, since genes are systematically transmitted from parents to children; for the same reason, what is heritable seems to be, of necessity, governed by the genetic patrimony. In reality, this equivalence is dangerously incorrect.

A trait is "heritable" when a certain similarity is observed between parents and children or, more generally, between sufficiently closely-related individuals. This concept, therefore, concerns only the level of phenotypes, that is, what can be directly seen and felt. As for the qualifier "genetic," it usually has no meaning at all: all traits are genetic, since they can only be expressed in an individual who is himself the product of a particular genetic heritage; language and religion can therefore be described as genetic. For this term to have real meaning, it must be used very restrictively: for instance, a trait can be labeled "genetic" only if a link has been discovered between its various expressions and the presence in the biological heritage of certain associations of genes.

Even in this narrow sense, the term "genetic" is by no means equivalent to "heritable." André Langaney (1978) uses "The paradox of sex and fortune" to illustrate this point: each person's sex is rigorously determined by his genetic heritage but this trait is not in any sense heritable; as for fortune . . . Similarly, phenylketonuria is a "genetic" disease because it only affects individuals with two copies of a well-defined gene that is transmitted according to the exact same pattern as that specified by the Mendelian model. However, even in this extremely simple sense, the environment intervenes: an appropriate diet makes it possible to prevent the disease from being expressed. What is "genetic" is therefore not necessarily inescapable.

Of course, the two concepts "heritable" and "genetic" are not independent, but the link between them is neither simple nor clear. Mendel's laws, the chromosomal theory of inheritance, are highly satisfactory for explaining the mode of transmission between generations of an "elementary trait," that is, a trait governed by a single pair of genes. However, traits subject to such simple determinism are exceptional; usually, we are interested in continuous traits for which it is inconceivable that we will one day establish a direct link between the genetic heritage, or genotype, and the measurable and observable trait, or phenotype. In the case of such traits, the intervention of numerous pairs of genes has to be assumed; what is still more important, the effect of these genes is a function of the environment in which they are found. In these circumstances, it is not possible to proceed by trying to establish deterministic cause-effect relationships between genotype and phenotype. We are obliged to study empirically, solely through the observation of measurable traits, the mode of transmission from parents to children.

The difficulty we have just outlined has led researchers to define the concept of heritability in various ways, which correspond to totally different approaches to the problem. We will present three of them here, that of biometricians, that of population geneticists, and finally that which corresponds to the fundamental question: to what extent is the genetic heritage responsible for the expression of a trait?

Heritability as Defined by Biometricians

The concept of heritability was first defined by biometricians, that is, researchers who use measurements in the study of organisms. Consider a measurable trait such as height. The graph in figure 11 represents observations carried out on an imaginary population in which we measured to the nearest centimeter the height of a large number of women and men, all members of couples with children, and, for each couple, the height of one of their female children. Each family is represented by a point whose abscissa is the average height of the parents and whose ordinate is the height of their daughter. In this way, one gets a "cloud" of points, for which one can define a "center" C, a point with, as abscissa, the general average of the heights of the parents in the population under study and, as ordinate, the average height of the daughters.

Consider all the couples of a certain height X; their daughters' heights are more or less dispersed around an average; we now complete our cloud with the points A_x representing the average height of the daughters for a given parental height (they are marked with an X on figure 11). When this kind of graph is drawn based on data from an actual population, two observations can usually be made:

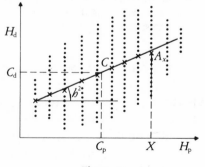

Figure 11

1. Deviations from the general average for the population are, on average, less in the case of daughters than in the case of parents. In other words, the difference between the ordinates of the points A_x and C is less than the difference between their abscissae: parents 10 cm taller than average have daughters who, on the whole, are only 8 cm taller than average. There is a return or "regression" towards the average.

2. The group of points A_x, representing the children's averages, are situated roughly on a straight line which is called the "regression line."

These are, to be sure, empirical findings which may very well not hold true for all cases. The most interesting aspect of this analysis is that it makes prediction possible. If one knows the height of the parents, one can estimate, before measuring it, their daughter's height with a degree of precision that increases as the dispersion of the cloud of points around the regression line decreases. It is easy to imagine how breeders can use graphs like this one: by selecting for cows with a high milk-yield, they can predict the average yield of their descendants. The response to this selection will be all the better as the regression toward the average is less, that is, as the slope of the regression line is steeper. This slope has been labelled "heritability" and it is usually represented by the symbol h^2.

— When this slope is equal to zero, the regression line is horizontal, the parental measure for the trait does not influence the average measure for the daughters, and the trait is therefore not heritable.
— When it is equal to unity, the regression line bisects the two axes, the deviation from the general average is the same in the children and in the parents, there is no regression towards the average, and the trait is strictly heritable.

This kind of research can, of course, be done on any measurable trait, whether it be head circumference, height, or annual income. If the points representing the averages for the parent-child couples are noticeably aligned, one is entitled to draw a regression line, measure its slope, and calculate the heritability of the trait. This estimate is based on data gathered from a particular population: it allows us to make a prediction about a child, based on information about its parents. However, it implies no hypothesis, it permits no deduction about the cause of the resemblance between parents and children. The entire argument was developed in a phenotypic context and does not allow us to make any inferences about genotypes.

Heritability as Defined by Geneticists

We know, however, that genes constitute the biological contribution of parents to their children; half of a father's genetic heritage is identical to that of his son; this degree of similarity between the genetic blueprints from which they have been developed naturally leads to a certain resemblance between them. Population geneticists, and in particular the famous English mathematician Fisher, have developed a theory for predicting this resemblance.

The procedure used is exactly opposite to that which we have just described. This time, we are working in a genotypic context: the mechanism of gene transmission is used to predict possible resemblances between phenotypes. To be sure, the influence of the environment, which was illustrated in figure 1, makes it impossible to go from genotype to phenotype unless we assume that "the environment is homogeneous." This assumption has to be accepted if the argument is to be continued. It is important to be fully aware of its significance: a genetic analysis of a quantitative trait has meaning only within a given context; it cannot be transposed to another context. Therefore, there are very narrow limits to the application of this theory. Agronomists are in little danger of forgetting this because they are in daily contact with the facts. In the next chapter, we shall see that some specialists in the human sciences, psychologists or sociologists, have, on the contrary, often carelessly transposed to their own domains the theoretical developments of geneticists, by analyzing certain results without due caution.

Starting from the fact that a trait is governed by a certain number of pairs of genes, Fisher proposed to look for the specific, individual effect of each gene on this trait.

Consider the simplest case, that of a quantitative trait T with only three possible values and governed by a single pair of genes A and a. If, in a given population, this trait has the following average values:

— 6 for individuals with the aa genotype
— 8 for individuals with the Aa genotype
— 10 for individuals with the AA genotype

it is clear that the a gene leads to a somewhat lower value for the trait and that the A gene leads to a somewhat higher value; these are, however, qualitative tendencies which we would like to pin down precisely by estimating the specific effect of each gene on the expression of the trait. A first obvious and essential point, but one which is often forgotten, is that this estimate depends necessarily on the frequencies, in the population, of the various genotypes. This point is illustrated by the following examples:

— Suppose that the genes A and a are represented equally in the population: they both therefore have a frequency of ½. The Hardy-Weinberg rule, explained in chapter 2, tells us that in this case the proportions of the 3 genotypes are: ¼ for AA, ½ for Aa, ¼ for aa. The average, T, is therefore 8. The effect of each gene can be explained by saying that the gene a lessens the value of T by 1 unit; and that A increases it by the same amount. Therefore, heterozygotes (Aa) are subject to an influence equal to $+1-1=0$, homozygotes (aa) $-1-1=-2$, homozygotes (AA) $+1+1=+2$, which corresponds to the reality as observed.
— Now, suppose that in a neighboring population where the gene frequencies are ¼ for a and ¾ for A, that is ¹⁄₁₆, ⁶⁄₁₆, ⁹⁄₁₆ for each of the genotypes, the average for the trait is $^{144}/_{16}=9$, a higher value than that of the hetrozygote. One can again ascribe a specific effect to each gene, but this time, we notice that the a gene causes a reduction of 1.5 and the A gene an increase of 0.5

This simple example proves two points:
1. The effect to be ascribed to each gene depends on the frequency of the genes in the population; it does not signify an effect

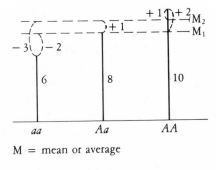

M = mean or average

Figure 12

proper to this gene, corresponding to an isolated determinism; it signifies an average effect in a certain population.

2. The effect ascribed to a gene does not concern the trait in itself but the difference between the average and the value of the trait for the various genotypes. It is not a matter of explaining the value found for the trait but the variations of this trait according to the genotypes.

This last remark is crucial. Strictly speaking, we cannot assert that a trait is governed by 1, 2, or n pairs of genes, but we can say that the variations of the trait are governed by a certain number of pairs of genes. This is by no means an insignificant nuance; it concerns our interpretations of genetic analysis in terms of "determinism" or "causation"; a trait may be subject to a multiplicity of determinisms involving a very large number of genes, but in a given population it may only show variations due to a single pair of genes. Genetic analysis will bring to light only the role of this pair of genes, which does not in any way take from the role of all the others in the determination of the trait.

Though conscious of these limitations, we can be reasonably satisfied; we were able, in the case under study, to ascribe an effect of its own to each gene and to observe that the trait corresponding to each genotype resulted from the sum of the effects of each of the genes present.

However, such equivalence is by no means the rule. Our analysis led to so simple a result only because of a particular property of the values that we had assumed: the difference between the *AA* and *Aa* genotypes was equal to the difference between the *Aa* and *aa* genotypes. When this equality does not hold true, it is impossible to ascribe an additive effect in this way to each gene. To circumvent this difficulty, R. Fisher proposed that the differences between genotypes be analyzed in two parts, one ascribable to the additive effects of the genes, the other to a "residue." For the model thus elaborated to have the greatest explanatory power compatible with the facts, an attempt is made to minimize these "residues," something which, with classic mathematical procedures (cancellation of the partial derivatives), is relatively easy. Thanks to this natural and efficient but totally arbitrary procedure, we can estimate the additive effects of the genes on the one hand and the "residual effects" corresponding to the interactions between them on the other.

Consider the preceding example once again, but assuming that, in a particular population, the trait under study has:

— a value of 6 for the group of individuals with the *aa* genotype
— a value of 10 for the group of individuals with the *Aa* genotype,
— a value of 8 for the group of individuals with the *AA* genotype.

The heterozygotes have a higher value, as is often the case in reality. This time, it is not obvious that a particular gene increases the value of the trait and that another lessens it. This classification can even be reversed depending on the frequencies of the two genes (figure 13):

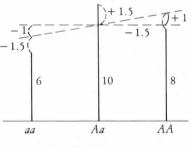

Figure 13

— in a population where the genes have equal frequencies, the average for the trait is 8.5. Fisher's method leads us to estimate that the gene *a* lessens the trait by 0.5 and that the gene *A* increases it by 0.5 (which leaves "residues," unexplained by the individual activity of each gene, of −1.5, −1.5, and +1.5 for each genotype; these residual effects are greater than the effects ascribed directly to each gene, but it is impossible to avoid such a paradox in this case);

— in a population where the gene *a* has a frequency of ¼ and *A* a frequency of ¾, the individual effect of the first is to increase the trait by 0.4, that of the second to lessen it by 0.1.

The roles ascribed to each gene are therefore reversed, but not due to any change in the biological mechanisms to which they are linked.

In fact, the principal significance of this analysis lies less in the calculation of the additive effects of the genes than in the establishment of a link between these effects and heritability as defined by biometricians. It can in fact be shown that:

"When the variations between the manifestations of a trait in different individuals can be analysed as being due in part to the differences between their genotypes, and in part to differences between their environments, and when these differences are independent, the slope of the parent-offspring 'regression line,' that is the heritability of the trait, is equal to the ratio between the variance of the additive effects of the genes involved and the total variance of the trait." (Remember that variance signifies the dispersion of a group of numbers; according to this formula, the heritability of a trait is the proportion of the total variation of that trait that is explicable solely by the additive effects of the genes involved.)

Because of this relationship, the two approaches, that of biometricians who observe similarities and that of geneticists who develop explanatory models, come together and complement each other. It is easy to understand how this unexpected outcome came to be viewed as proof of the solidity and appropriateness of the conceptual framework used. However, there is a danger of forgetting the price that had to be paid for this success; to reconcile the two points of view, it was necessary to make some very important simplifying assumptions about the noninteraction and the independence of the genetic and environmental factors.

The concept of heritability, thus defined as the relationship between the variance due to additive effects and the total variance, is limited in its meaning and application and is best described by the American term "heritability in the narrow sense," which can be represented by the symbol h_n^2.

Since it measures the individual effects of each of the various genes independently of the other genes and of the environment, "heritability in the narrow sense" is very useful in the development of methods for species improvement. It permits the choice of the most efficient selection techniques, based either on individual performances (if h_n^2 is high), or on average family performances (if h_n^2 is low). Note, however, that it does not in any sense constitute a measure of the importance of the genetic heritage in the determinism of the phenotype. A trait with a strictly genetic basis may very well have a heritability of zero. Such would be the case, for instance, for the trait described earlier (where the three genotypes lead to the values 6, 10, and 8 respectively) in a population where the frequencies of the two genes were ⅓ for a and ⅔ for A. It is easy to prove that the child-parent "regression line" is horizontal in these circumstances; on average, there is no phenotypic similarity between fathers and sons even though, according to our hypothesis, the differences between individuals were due exclusively to the differences in their genotypes.

Finally, remember that the parameter h_n^2 cannot be estimated directly, based on the additive effects of genes, since these cannot be observed. The only possible method is to compare similarities between individuals with differing degrees of relatedness (son-father, half-brothers, brothers, cousins . . .) and to estimate h_n^2 as a function of the differences that the theoretical models lead one to predict.

This estimate assumes that one was successful in eliminating the possibility of a correlation between genotypes and environments, which can be done in agriculture through carefully planned experiments, but which cannot be contemplated for the human species.

In fact, "heritability in the narrow sense" has no application in the study of human traits. Only "heritability in the broad sense" can be used. This is what we are now going to define.

Heritability as Defined by "Those Who Are Interested in the Degree to Which the Genome Contributes to the Expression of a Trait"

Neither of the two definitions of heritability we have just outlined allows us to answer the question that springs spontaneously to mind when we study a trait that is, to all appearances, subject to the influence of both genes and environment. This question is: to what extent are differences between individuals due to the genotype on the one hand and to the environment on the other?

To this question there is a classical answer, supplied by the mathematical technique called "analysis of variance." The principle underlying this technique is simple. Consider a trait with, in a particular population, a certain dispersion characterized by its variance. First, group together all individuals with the same genotype: the differences between them are entirely due to the differences between their environments, the residual variance V_e measures this influence. Next, group together all the individuals living in the same environment, the differences between these are due to the differences between their genomes, the residual variance V_g measures this influence. Naturally, the total $V_e + V_g$ is usually not equal to the total variance V; a supplementary term, $I(GE)$ must be introduced to make the equation:

$$V = V_g + V_e + I(GE)$$

The term $I(GE)$ signifies the nonadditivity of the partial variances; it can be seen as a parameter measuring the *interaction* between the environment and the genotype.

In some cases, it appears that the corrective term $I(GE)$ is of little importance compared to the terms V_g and V_e. It is natural, in such cases, to consider that our equation represents an analysis of the differences between individuals into components corresponding to the effects of the environment and of the genotype. The effects of the genotype can be written in the form V_g/V and it is often assumed that this measures, at least approximately, the heritability of a trait. To underline the fact that this is a new definition of the concept, it is useful to refer to it as "heritability in the broad sense," represented by the symbol h_b^2 or H^2.

In human genetics this parameter h_b^2 has been widely used because it can be estimated directly from studies of twins. "Monozygotic" or "true" twins, that is, twins resulting from the fertilization of a single egg by a single sperm, are two individuals with exactly the same genetic heritage; the contrary, "false" or "dizygotic" twins, are simply brothers or sisters conceived simultaneously. One can see how, by analyzing the differences between the members of a pair of twins, depending on the type to which they belong, we can estimate the proportions V_g and V_e of the total variance. Here again, it is necessary to check that numerous conditions are satisfied, in particular, that the twins in question are subject to as great a dispersion of environmental influences as the population as a whole. In the next chapter, which deals with the problems raised by psychologists' research on the "heritability of intelligence," we will see that these conditions are in fact very rarely satisfied.

For the present, let us outline more clearly the meaning of the initial hypothesis which affects the entire argument: the term of interaction $I(GE)$ is assumed to be neglible. To understand this assumption more clearly, consider an extremely simple case: that of a population where the trait under study is such that the environment can be represented by a single number E (this would be the case, for instance, if the only variable parameter influencing the trait were elevation, or temperature, or the amount of food available to each individual) and where only 2 genotypes are present.

The value of the trait T in an individual is a function of the value of E for the environment and of his genotype. Suppose that for each of the two genotypes, the coordinate points (T, E) are on a straight line and that the two lines are parallel. The value T_i of the trait in the individual I can be analyzed, in that case, in three parts:

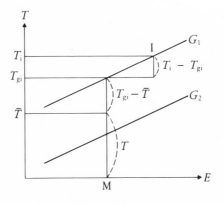

Figure 14

— the mean value $\bar{T}$ for the whole population
— a difference $T_{gi} - \bar{T}$ between the mean for the trait in the group of individuals with the g_i genotype, and the overall mean $\bar{T}$
— a difference $T_i - T_{gi}$ due to the fact that I is not subject to the average environment.

The second term represents the effect of the genome, the third, the effect of the environment. The fact that the two lines are parallel indicates that a given environmental variation causes the same varation in the trait, regardless of genotype, and that the difference between two individuals differing only in genotype is the same regardless of the environment. In these circumstances, we can write:

$$V = V_g + V_e$$

and the term $h_b^2 = V_g/V$ has a clear meaning: it represents the proportion of the total variability that is due to genetic differences. If one wishes, one can see it as an indicator of the genetic determinism of the differences in the expression of this trait in the population.

If, on the contrary, our two lines had not been parallel, the sum of the two partial variances would not have been equal to the total variance; the term $I(GE)$ would have had to be introduced and that would have deprived the parameter h_b^2 of a large part of its significance.

Analysis of Variance and the Analysis of Causes

The term "indicator of genetic determinism" is doubly misleading. It gives the illusion that we have analyzed the determinism of a trait into its independent causes. In fact, our study dealt not with the *trait* in itself, but with the differences observed; it investigated not the *causes* of these differences, but the way in which they varied when certain factors were fixed.

It is realistic to interpret our results in terms of "causes" only if the underlying mechanism is determined by factors acting independently of each other, and adding up their effects.

To illustrate this point, consider an example: two categories of masons, let us say Bretons and Jurassians, are laying bricks on top of each other to make a wall; I can count the rows and estimate how much each category contributed to the total height. After a year of observation, suppose I find that the Bretons have laid 80 percent of the bricks and the Jurassians 20 percent; I will therefore be able to say that the final result is determined 80 percent by the former and 20 percent by the later and I really will be able to analyze the causes of this result. However, if the Bretons are responsible for making the cement, and the Jurassians for laying and sealing the bricks, this analysis no longer holds. It would be absurd to try to estimate the percentage of the final result which was determined by each group since it was their interaction alone that was efficacious; neither the Bretons nor the Jurassians built even the tiniest portion of the wall on their own.

However, during the year in which the study was done, absenteeism affected both categories, leading to a certain variation in the daily production, a variation that can be measured by the variance V of the wall surface completed each day. I can group all the days when the number of Bretons at work was the same and calculate the variance of this surface; it is due to fluctuations in the number of Jurassians, let us call it V_J; similarly, I can estimate V_B, the variance in production calculated from all the days when the number of Jurassians was constant. If, by chance, the total $V_B + V_J$ is close to V, I can conclude that V_B/V represents the portion of the total variability that is due to the variability in the number of Bre-

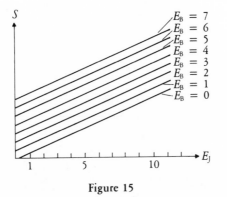

Figure 15

tons, and, likewise, that V_J/V represents the influence of the Jurassians on the variations in daily production.

Let us draw a graph similar to that in figure 14, with, as ordinate, S, the amount of wall surface built each day, and as abscissa, E_J, the number of Jurassian workers present; we obtain a series of points situated on curves which depend on the number E_B of Breton workers present. If there is no interaction between the two categories (for instance, if each worker, regardless of his category, lays rows of bricks) the graph will consist of parallel straight lines (figure 15). However, if their activities are interdependent (for example, if the former work the concrete mixer while the latter lay and seal the bricks), the curves are highly unlikely to be parallel straight lines. In this case, threshold phenomena will appear: if there is no Breton at work, the Jurassians, regardless of their number, will be unable to build anything, for lack of cement; if there is only one Breton, the amount done will be very small, etc.; our graph will be similar to that in figure 16. First of all, note that, in this case, the analysis of the variance leads to different results depending on the range of variation in the number of Jurassians. On a building site A, where the number of Jurassians varied between 1 and 4, we note that the total variance is due essentially to V_J and only slightly to V_B. On a site B, where they varied between 8 and 12, in a zone where a threshold effect appears, the variance V_J is almost zero, V being represented essentially by V_B. Even in such a simple case, the results of a variance analysis can be contradic-

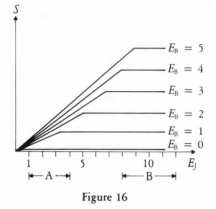

Figure 16

tory from one observation to the next; in no case can they give the least indication of the underlying mechanism.

Unfortunately, this mathematical technique, which is perfectly legitimate in many cases, is used by practitioners of many disciplines, who sometimes interpret the results in a totally inappropriate way. When a phenomenon is quantified, it is always possible to apply complex mathematical analyses to the measures obtained, and to eventually estimate various parameters. However, if these parameters have no precise meaning, the calculations leading up to them are a complete waste of time, even if subtle algorithms and powerful computers were used. An absurd question remains absurd, regardless of the complexity of the mathematical techniques used to answer it. This is often true of attempts to estimate the contribution of various "causes" to the determinism of a trait. In all but exceptional cases, these attempts are prompted by a totally meaningless question.

This distinction between the search for causes and the analysis of variations is often forgotten in studying the problem that concerns us here, that of estimating the exact role of the environment and of the genetic heritage in the determinism of a trait. We will illustrate this difficulty with a very simple example taken from agricultural research.

One Real Case Among Many: The Yield of Beans

It was while studying peas that Mendel discovered the laws of genetics. We will use a closely related species, beans, to illustrate the nonsense of some of the claims made about the role of the genome in the determinism of certain traits, or in the determinism of their variations.

Primitive varieties of beans, such as those traditionally grown in South America, have a relatively low yield, varying, of course, according to the quality of the soil and the amount of manure added. In the barren and sparse soil of some Mexican plateaus, of Peru, or of Haiti, the yield is barely more than 5 quintals per hectare. The same seeds can yield from 10 to 15 quintals per hectare in the agronomical research stations of those countries. Some of these varieties produce the same yield of 15 quintals in good soil in France. However, since the introduction of this species to Europe in the sixteenth century, breeders have been applying their "improvement" techniques to create new varieties that are better adapted to France's climate and to the richness of its soil. Nowadays in France, it is by no means rare to reach a yield of 30 quintals per hectare, six times higher than that of Haitian peasants. This difference has two "causes": the genetic difference between the varieties and the differences in growing conditions. This prompts the question: how much of the difference in yield is to be attributed to each of these causes?

In replying, one can point out that changing the environment of the Indian variety increases the yield by 10 quintals, and that genetically changing the variety increases the yield on French soil by a further 15 quintals per hectare, hence the conclusion: the improvement is due 40 percent to the change in environment and 60 percent to genetic change.

However we could, for equally good reasons, take the opposite course and plant the "improved" European variety in Haiti or in Peru. It so happens that this variety is unable to survive in the poorer soils of those countries. Its yield there is zero. We can therefore claim that the difference between yields corresponds to

+30 qs due to environmental differences and −5 qs due to genetic differences.

Variance analysis therefore has no absolute meaning: the improved species performs better than the traditional one in a certain environment, and worse than it in other environments. This finding is represented by the graph in figure 17: this time, the two curves representing the variation of the trait "yield" as a function of "environment" are far from being parallel straight lines, they are lines which intersect. In a case like this, the question we had asked ("How much of the total difference is due to the various causes?") has no meaning. This is by no means an exceptional example, maliciously chosen just to create confusion. In a great number of cases, the interaction between genotype and environment is such that the trait under study does not allow one to classify the genotypes.

Note that the analysis of variance can, however, be useful, provided the environmental variations are sufficiently small. French agronomists, working within the F range of environmental variation, are in a position to evaluate a "heritability" for the yield, just as Peruvian agronomists can by working within the P range of variation. Their results will be perfectly valid and will allow them to intervene effectively provided they do not stray too far from their own domain of variation. The French can, for instance, claim that the bean-yield depends 20 percent on the genetic heritage and 80

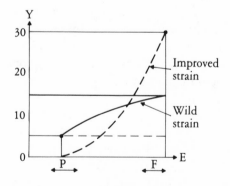

Figure 17

percent on the environment, the Peruvians can say that these figures are 80 percent and 20 percent respectively, and they will both be correct. These results will not be consistent with each other because they will be referring to two different situations: one will represent the heritability of yields in environment F, the other in environment P; neither will represent "heritability in itself," a concept that cannot be defined. If one were to use these partial results to try to explain the overall differences between the French and Indian yields, one would obviously be engaging in an absurd exercise. Agronomists are well aware of this and would not waste time on such foolishness.

Even at the risk of boring my readers, I think it useful to repeat: a mathematical presentation cannot confer meaning on an inept measure. A disturbed geneticist or an insane psychologist may one day invent the parameter X obtained, for each head of a family, by dividing his height by the head-circumference of his spouse and by adding the mean of their children's intelligence quotients; they may give X a Greek name or, better still, an English one, calculate X in a large number of families, compare the means obtained for X according to socioprofessional groups, races, or generations, determine the heritability of X, etc. The profusion of calculations will not prevent the results from being utterly meaningless, since they are based on numbers that measure nothing.

How many quarrels could have been avoided if, before bringing numbers into a discussion, people were willing to honestly question their meaning.

This is what we will attempt in the next chapter which deals with the trait that is at the center of all eugenic thinking on the possibility of improving the human species: intelligence.

Questions and Doubts

Genetic models dealing with quantitative traits do, to be sure, allow selective breeders to plan their breeding programs, but one must not allow oneself to be misled as to their explanatory power. Unrealistic by definition, they should be seen more as a method for solving the numerical problems with which researchers have to deal than as an attempt at a description of a natural mechanism. Theoreticians gradually began to feel somewhat uneasy; in 1976, one of the more famous among them, Oscar Kempthorne, author of a work well known to all geneticists, thought it useful to call a meeting of about fifty European and American geneticists; it took place at Kempthorne's own University of Iowa, which is surrounded by vast corn fields, tangible proof of the effectiveness of the selective methods. The purpose of the meeting was to reflect on "the profound doubts felt about the quality and usefulness of the theories currently available in quantitative genetics." A book almost 900 pages long was published after this meeting; it shows the extent to which any further progress is dependent on the elaboration of models that are less simplistic than those used up to now (Kempthorne et al. 1977).

The current techniques, especially those that are based on the concept of heritability, involve the analysis of small local variations; this approach is equivalent to that of the mathematician who reduces a curve to its tangent, or who simplifies the values of a function to that of the first term of its expansion in a series. Usually, this is the only possible approach, because of lack of information or of techniques that would enable one to "stick more closely to reality." However, one must have no illusions either about the significance of the parameters being calculated or about the validity of a result that requires repeated applications of techniques that are sure to be accurate only when applied in a single-step manner.

At each stage of their efforts, selective breeders have, for instance, improved strains of beans; there is no doubt but that each successful step is interesting, but is the final result really an improvement? The new strains are wonderfully productive under the very special conditions that we can provide for them; they are incapable of surviving the conditions usually provided by a natural environment. Is the genetic heritage of these strains better than the ancestral heritage? Or is it worse? This question cannot be answered. The result depends on the environments in which we make the comparison.

What is true of beans is also true of most of the other species that we have manipulated, domesticated, and put at our service. If the seeds that are now used to grow splendid wheat in La Beauce had been available to our farming ancestors two centuries ago, they would have yielded very poor harvests, much inferior to those yielded by the traditional seeds. The extreme case is no doubt that of corn, which had been selected by the Maya Indians for several thousand years, before geneticists speeded up the process; the varieties currently grown are so far from having the traits required naturally for reproduction that they cannot be perpetuated without human intervention. If a biological or atomic cataclysm were to destroy humanity, corn would also become extinct. Only one species of corn would survive, téosinte, which is considered a weed, and which seems to be a distant ancestor of corn or at least another related descendant from a common ancestor. As for animal species, many of them have reached a stage of specialization that makes them dependent on humans for survival; race horses can, to be sure, run remarkably fast, but they are only poor frail animals, unable to cope on their own with the slightest attack from the environment.

The very term "species improvement" is misleading. We have improved neither corn, nor cows, nor horses; what we have improved is corn's capacity for benefiting from fertilizers, cows' capacity for producing milk, and horses' capacity for racing. Once again, the words that we are using signify something different from the reality and, through a series of slips, they have reached a point where they signify the opposite to that reality. Can we boast of having improved corn or horses, when we have made them into species incapable of living without us?

SEVEN

Intelligence and Genetic Heritage

MOST SOCIETIES fear biological decadence or even "degeneration." The variability of individuals, the similarity between children and parents lead inevitably to this conclusion: for the good of the group, the "best" should contribute more than the others to the transmission of the biological heritage. When dealing with arguments about races and their evolution, we quoted a sentence from Konrad Lorenz that is typical of this way of thinking. The "casual" remarks that we all hear occasionally are revealing: this attitude is shared by the majority. In the face of this almost general consensus, one should neither agree out of conformism nor disagree for the sake of being original; one should reflect on the meaning of this spontaneous eugenism.

The first question to be asked is, of course: what does "best" mean? This term can indeed have a precise meaning, but one that will, of necessity, vary depending on the society, or on the pressures exerted on the group by the outside world. In a hunting tribe, the "best" are those with the most penetrating eyesight, the most agility, and the quickest reflexes; in an agricultural tribe, those with the most perseverance, who know best how to organize their work, adjust to the rhythm of the seasons and prepare the harvests to come; in our supposedly "advanced" societies, certain physical qualities are highly prized; the infatuation with champions from every discipline is proof of this; but one "quality" seems to win over all others, intelligence.

The Superman of the comic strips does have strong arms, but, above all, he has a great brain. Implicitly or otherwise, all programs for the improvement of humanity aim at creating beings of superior intelligence. What exactly is this quality which all agree is essential?

What Is Intelligence?

A word's ability to communicate precise information is diminished if it has a wealth of different meanings. The Oxford dictionary has several columns devoted to intelligence, so varied are the concepts it signifies; when using this word, it is therefore necessary to always indicate clearly what meaning should be assigned to it.

The characteristics evoked by the word "intelligence" are rather poorly defined, but at first sight they seem specific to our species. Humanity has long considered intelligence to be that "special something which differentiates a human being from other beings" (J. -P. Richard 1973); note that these "beings" can be computers just as much as animals.

Philosophers and psychologists have, to be sure, tried to pin down this special something in very "intelligent" treatises. They are almost unanimous in considering intelligence to be an ensemble of abilities, a kind of power or energy, whose nature is (and doubtless always will be) unknown to us, but which manifests itself in certain ways. A capacity for abstraction or for imagining appropriate behavior in a new situation thus seem to be important facets of intelligence. However, these capacities are by no means exclusive to humans. They are found, in certain forms, in animals: any kind of behavior, no matter how rudimentary, requires some capacity for abstraction. For many psychologists, intelligence is not a specifically human property, but a set of properties that are especially well developed in our species. Criteria of "all or nothing" or of "presence or absence" are less relevant to it than are quantifiable parameters specifying "more or less."

One is thus led to see intelligence as a set of quantitative, rather than of qualitative, traits, which can therefore be measured. All the arguments about intelligence deal, in fact, not with this entity which is so difficult to narrow down and define, but with measurable parameters which are assumed to be representations of this entity. To illustrate this difference, consider the image used by Wechsler, a psychologist who created many universally accepted tests (quoted by P. Dague 1977): we attribute to a certain "something," which we call *electricity*, the heat produced in a wire when an electric current passes through it; we can characterize this unknown reality according to the number of calories generated by it under certain conditions. We are therefore no longer speculating about this reality itself, we are merely observing what it produces. We can make measures, calculations, and comparisons: the reality we are studying has become "scientific." However, this success may be dearly bought: measuring the amount of heat gives us no hint of other and stranger properties of electricity, magnetic or chemical ones for instance. All we are doing is studying a parameter arbitrarily chosen to represent an inaccessible reality.

Such is the role of the famous intelligence quotient (IQ) which is so frequently used to represent intelligence.

Mental Age and Intelligence Quotient

Since we are unable to specify what intelligence is, we must content ourselves with characterizing it according to an individual's level of proficiency in certain skills (judging, understanding, imagining, etc.); these skills can be measured from attitudes that are revealed by the replies given to questions in "tests" developed by psychologists with a view to triggering certain relatively well-defined mental operations.

Psychologists in this field have shown remarkable subtlety and imagination. They have invented a wide range of tests which, from several angles, give different glimpses of that elusive and ever-changing "object," intellectual activity. However, precisely because of their multiplicity, the data thus gathered present a problem: how are they to be synthesized? The ingenious solution proposed by Binet at the beginning of this century is based on the observation of intellectual development in the child. In a given population, a particular test is passed, on average, by children over a certain age, but not by children under that age. In this way, it is possible to establish a scale indicating the age at which each test can usually be passed. The results obtained by a child in a set of tests can be used to calculate his "mental age" in years and months; by comparing this with his actual age, one can specify his position in relation to the average, and how far behind or ahead he may be.

This procedure is convincing especially because of the concrete meaning it gives to the final result. Instead of calculating an average score unrelated to experience, an age is determined, and our mind grasps this quite readily.

However, this achievement must not lead us to forget that, like any average, mental age is a single piece of information which retains only a tiny fraction of the information contained in the initial results. Two children with very different results for the individual tests may, due to the effects of compensation, have identical mental ages. The vision that this number gives us of a child's overall intellectual abilities is, therefore, very inadequate.

Moreover, the significance of a difference between actual age and mental age is only relative: an advance of one year at fifteen has far less importance than the same advance at five years. To eliminate this objection, American psychologists Stern and Terman proposed the concept of "intellectual development quotient," or IQ, which has since become world-famous. To justify this IQ, they assume the following hypothesis: each child's intellectual development proceeds in a continuous manner and at a constant rate. The points representing this development on a graph, with the actual age as abscissa and the mental age as ordinate, are therefore on a straight line.

The slope of this line characterizes the speed of this development, it measures the IQ; more precisely, the IQ is equal to this slope multiplied by 100, which gives more manageable numbers: the IQ is therefore, by definition, equal to 100 times the quotient $\frac{\text{Mental Age}}{\text{Actual Age}}$. Children whose mental age is the same as their actual age

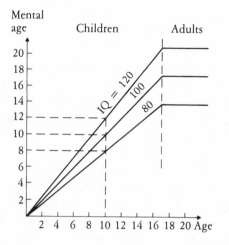

Figure 18

age are on the line IQ = 100; those who, at 10 years, have a mental age of 8, or at 15 years, a mental age of 12, are on the line IQ = 80, and so on.

If, hypothesizing further, we assume that this uniform development ceases at the same age for all individuals, then the IQ as measured in children also represents the intellectual hierarchy of adults.

It is pointless to criticize psychologists for having thus accumulated hypotheses of questionable accuracy; any scientific argument must, of necessity, proceed in this way. These hypotheses must, however, be explicit and frequently restated to prevent the results obtained from gradually acquiring in discourse an absolute meaning that they are far from having in reality.

The Instability of IQ

For IQ to be considered a stable characteristic in an individual, a kind of label that can be assigned to him for once and for all, both of the hypotheses that we mentioned must be verified: if changes are observed in the slope, the hierarchies may be upset; if developmental times vary, the hierarchy among adults no longer reflects that among children. Psychologists know well that these changes are frequent. Longitudinal studies, comparing the IQs of the same individuals over time, have attempted to establish whether IQ is stable. However, such studies are difficult and costly. The results currently available are far from being coherent: some indicate a good degree of stability, but most show considerable changes. W. Bodmer (1976:505) cites a study carried out in Sweden in which the average for a group of students had increased by 11 points between the age of 12 and the end of their studies.

Our current understanding of intellectual development, particularly since Piaget's work, has little compatibility with the hypothesis of continuous progression over time. For Piaget, this development proceeds through successive acquisitions of new functions in a specific order, but with variable age intervals according to the individual. At each stage, the mind integrates its previous capacities into a richer whole, which will, in its turn, provide some of the elements necessary to a new integration. It is much more a matter of reaching successive thresholds than of following an even path to a single peak. In order to specify intellectual possibilities at a particular age, it is therefore appropriate to assess what stage the child has reached. This assessment is more qualitative than quantitative (Loehlin et al. 1975:280–285). This concept of intelligence is not, of course, compatible with IQ measures, whose significance can no longer even be defined.

Moreover, a change of environment can noticeably change intellectual development, as psychologists know quite well. Proof of this comes especially from certain results, analyzed by D. Courgeau (1973), from a huge study carried out by the Institut National d'Etudes Démographiques on the intellectual level of school-age children. Of the children studied, 2,631 were born outside of France, mostly in Italy, Spain, Portugal, and North Africa. The IQs of these children were analyzed as a function of the length of their stay in France; the result was remarkably clear:

— the new arrivals have lower IQs than their compatriots who have been in France for a long time,
— this difference is gradually and regularly eliminated during the first five years of their stay,

— the average IQ thus increases by 10 points in four years, and this increase is the same, regardless of the country from which the children had come.

It is difficult to imagine clearer proof of the impossibility of using IQ as a definitive label for each individual; individual experience can modify it considerably. IQ is a measure that reflects a certain phase in each person's development, and this development depends very largely on his or her life experiences.

The Imprecision of IQ

The fact of being observed sometimes changes to a small but not insignificant degree the object under study. This is particularly true of IQ measures: the interaction between the observer and the person observed is such that the result is strongly influenced by the psychologist's behavior, by what he knows, or even by what he expects. Professor Robert Rosenthal of Harvard did many studies on this point and brought to light what he called the "Pygmalion effect" (1971). According to his studies, the simple fact of telling a teacher that a particular child has a high "potential for intellectual achievement" is enough to cause an increase in the student's IQ during the school year. Furthermore, students whose performances measure up to expectations in this manner are considered nicer and more interesting than students who succeed in spite of the psychologist's prediction (Rosenthal 1974). Intellectual development may therefore proceed under more favorable circumstances when it is expected; the chances of achievement may be increased by the prediction, which thus becomes a self-fulfilling prophesy. Rosenthal's results have, however, been contested by many psychologists who, in similar experiments, were unable to find such marked differences.

The effect of the observer's attitude and personality on the result of the tests has often been noted. Depending on whether he is severe or amiable, white-skinned or dark-skinned, male or female, attentive or seemingly indifferent, the result may vary considerably. In his *Manuel pour l'examen psychologique de l'enfant*

(1960), R. Zazzo describes the results of his experiments on this subject. A. Jensen, for his part, notes that underprivileged children are often inhibited during the first tests to which they are subjected and are in fact more intelligent than the results suggest. By creating a trusting and relaxed atmosphere, he found "as a general rule, an increase of 8 to 10 points in IQ."

Whatever their causes, the possible errors in the estimates of IQ should be systematically allowed for in the various arguments concerning this parameter. In all the sciences, the estimate of possible errors is part of metrology. It is valid to use a result only if one indicates the degree of precision with which it has been determined: a physicist knows quite well, for instance, that mass cannot be calculated with rigorous accuracy. The imperfection of the equipment available or of the conditions under which the observation is carried out inevitably causes a degree of uncertainty which can itself be estimated. It is customary to specify a "confidence interval," a range within which the actual measure has 95 chances out of 100 of being accurate. A physicist claims not that "a particular object has a mass of 150 g," but "a mass of 150 ± 2 g." In other words, the measure is indeed 150 g, but, given the imprecision of the equipment or of the techniques used, all we can say is that the "true" mass is, with a probability of 95 percent, between 148 and 152 g.

IQ measures are not, of course, free of this uncertainty. However, it is rare for psychologists themselves to admit to the problem. For a scientist, this omission is enough to make the measure suspect. The "human sciences," which lay claim to the status of "exact sciences," must pay the price: a minimum of rigor in definitions, of discipline in measurements.

The tests used to measure IQ have been standardized very carefully. No one, however, would go so far as to claim that they lead to an "exact" measure. The concept of a "confidence interval" is, of course, less clear-cut in the case of IQ than of physical size, since the "true" IQ cannot be defined. In the case of IQ, the best one can do is to indicate the dispersion of the various values derived for the same individual from independent IQ measures. Consider the figures published on this subject by psychologist Pierre Dague

(1977). According to him, the confidence interval is in the order of ±5 for IQs lower than 85, ±10 for IQs between 85 and 115. In other words, the statement: "Lucien's IQ is equal to 105" means: "Independent measures of Lucien's IQ would, 95 times out of 100, lead to a result somewhere between 95 and 115, and 5 times out of 100 to a result outside this interval."

The second proposition is, of course, less precise than the first. Nonetheless, it alone gives an honest interpretation of the result of the tests. The vagueness of the measure obtained corresponds to the facts; camouflaging it can only lead to inexact interpretations.

The situation is similar when two IQs are being compared: if Lucien's is 105 and André's 102, the difference between them, +3, is also a measure with a confidence interval. This interval is even greater than that for each of the individual measures, it is 14 points; in other words, the difference between Lucien and André is somewhere in the −11, +17 range. This result is so imprecise that it is unlikely to be useful. It does not even allow one to say which of the two children has a higher IQ: the probability that Lucien's IQ is higher than André's is, in this example, only 72 percent, which means that there is a probability of 28 percent that André's is the higher of the two. In order to be able to say with a less than 5 percent risk of error that Lucien has an IQ higher than André's, the difference between the measures must be more than 8 points; in our example, this statement is possible if André's IQ is lower than 97.

Finally, do not forget that IQ is defined as a quotient, which limits the meaning of arithmetic operations based on this measure. In particular, the sum of two IQs represents nothing, the average of the two precious little. IQ is not a measure comparable to that of a mass or a length; it is only a marker, relative to a scale of reference. To be sure, this is also true of the various physical measures, temperature for instance. However, in the case of IQ, this scale of reference is itself based on the distribution of measures in an arbitrarily chosen population; it cannot be anchored to any stable phenomena found in nature, such as the boiling point of water for determining the 100° mark in the temperature scale.

The figure obtained for IQ merely allows one to situate each individual in relation to the group of individuals in the reference population. If someone has a score of 100, he is equalled or surpassed by half of this population, if he reaches 115 by only 16 percent, but if he obtains only 85 he is surpassed by 84 percent. It would no doubt be less misleading to indicate these percentages, instead of proceeding as though IQ represented an absolute measure.[1]

What Is the Purpose of IQ?

It was at the request of the Ministry of Education that Binet designed sets of tests for measuring mental age. The aim was to detect the risk of failure at school. Tests were adopted as a function of their correlation with the results achieved by the children in their studies. Successive adjustments made it possible to select combinations of tests that fulfill this function very well. IQ is, at least on average, a good indicator of a child's chances of success, or risk of failure, in school.

IQ is therefore useful above all for forecasting, and this is by no means insignificant. However, this objective is entirely different from that of diagnostic tests. Moreover, in societies like ours where "success" throughout one's professional life is very closely linked

1. This observation is important, especially for correctly interpreting the averages calculated from IQs. Consider two families, A and B, each composed of two children. The IQs are 108 and 110 in family A, 100 and 120 in family B. Which family has the higher average IQ? Naturally, the averages can be calculated as 109 for family A, 110 for family B, to reach the conclusion: B>A.

However, if one wished to adhere more closely to the actual significance of IQ, one could say that the children of A are surpassed respectively by 29.8 percent and 25.5 percent of individuals in the reference population, that is 27.6 percent on average, which corresponds to an IQ of 109.0. Similarly, in family B, these percentages are 50 percent and 9.3 percent, that is an average of 29.65 percent corresponding to an IQ of 108.0; hence the conclusion: B<A.

IQ does not, therefore, allow one to compare two families unambiguously. This is due to the impossibility of validly adding two IQs; an average cannot be calculated unless the numbers can be added.

These difficulties do not, as we shall see, prevent certain psychologists from comparing the averages for populations or social groups, without taking the slightest precaution.

to success at school, IQ is also a good tool for predicting the future chances of a child.

It is not surprising in these circumstances that the IQs of adults vary greatly according to socioprofessional category. This finding is not a discovery but is a result of the very definition of this parameter.

Remember that the first sets of tests proposed by Binet led to a higher IQ average for girls than for boys; changes were made so as to eliminate this difference. To now offer the equality of IQ averages as proof that "girls and boys are equally intelligent" is nonsense; this equality is simply a consequence of the definitions adopted. It is likewise nonsense to say that the IQ of teachers is higher than that of unskilled workers: this statement merely means that teachers succeed better at their studies than unskilled workers; do we need tests to confirm this obvious fact?

It would, however, be unfair to claim that IQ is of no use whatsoever: together with its usefulness for the detection of risk of failure at school, it has shown itself a useful guide for helping children with difficulties in school, but who nonetheless get good results in the tests. What is important is to be conscious of the limits of this measure, and to use it more as a basis for questions than as an answer. The incredible popularity of IQ is linked, no doubt, to people's infatuation with anything that is vaguely mathematical. In many cases, IQ is but a cover used by psychologists, who quantify the object of their study even before having defined it, or verified that it is definable. This weakness is especially obvious when some of them use genetic concepts and study "the heritability of intelligence."

IQ and Genetic Heritage

Here we come to the eternal and fundamental question: to what extent are we determined, to what extent free? Are we the inevitable product of the genetic heritage assembled by chance at the instant of our conception or the result of the human experiences that we have lived, endured, but also to some extent created? These questions have been so much debated over the centuries, in terms adapted to the thinking of each epoch, that there seems to be little hope of reaching a clear conclusion that would be as convincing as a scientific proposition. Three centuries ago, the position adopted by the Dutch bishop Jansénius on the "predestination of divine grace" triggered passionate debates which shook the Church; nine years ago, statements made by American psychologist Arthur Jensen about the "genetic determinism of intelligence" provoked a violent quarrel which is still going on in universities. The terms are different but the problem is similar: in one case, it is a question of God and of the soul's salvation, in the other of genes and social success; in both cases there is the question of submission to a destiny, or of taking charge of one's own becoming.

The big change comes from the role that is now attributed to science in this controversy. Most of the current statements on this subject begin with: "it is scientifically proven that . . ." and continue: ". . . intelligence is determined 80 percent by the genetic heritage and 20 percent by the environment" (Vianson-Ponté). This sentence has been repeated so often that it has acquired the status of an eternal truth. Strictly speaking, however, it has no meaning.

Note first of all that speaking about intelligence in terms of percentages implies that this trait is quantifiable. In fact, it is not a question of intelligence but of IQ, which is not at all the same thing. Above all, these percentages have meaning only if the two causes mentioned, heredity and environment, have independent and additive effects. I am correct in saying that the state's revenue for the coming year will depend 56 percent on direct taxes and 44 percent on indirect taxes, because everyone understands that the elimination of the former would reduce this revenue by 56 percent; my statement has meaning because, in this case, there is additivity.

The only meaning that the percentages for intelligence could possibly have is the following: a child who received no environmental input would have an IQ of 80, a child who received no genes would have an IQ of 20. These statements are so absurd that no one would dare make them; however, attempts at analyzing the determinism of intelligence are just as absurd.

In reality, these figures are based on studies of variances, studies which may be perfectly legitimate but which do not allow one, in any case, to evoke the notion of determinism. These studies are based on the concept of heritability, the difficulties and limits of which we discussed in chapter 6.

"Heritability as defined by biometricians," which measures the similarity between parents and children, cannot, of course, be applied to the study of IQ; it is too obvious that this similarity results from the combined influence of environment, education, and genetic relatedness. "Narrow-sense heritability," an estimate of the additive effects of the genes, cannot be used here, because there is no possibility of defining the various genes that directly influence the value of IQ or of carrying out the systematic crosses that allow agronomists to estimate this parameter indirectly. Only "broad-sense heritability" can be useful, for analyzing the observed differences in IQ (and not for analyzing IQ itself) into a fraction due to environmental differences and a fraction due to genetic differences.

To do this, we need to study individuals who are genetically identical but who are subject to different environments, that is "real" (monozygotic) twins reared separately. The problem here is to gather data that are extensive and precise enough to be meaningful. Twins occur in barely 1 percent of births, and only a third are monozygotic; experience shows that, of the latter, barely 1 pair per 1,000 are separated while still very young. It is therefore not surprising that studies of monozygotic twins reared apart are few and involve very small numbers: 19 pairs in the classic study carried out by Newman, Freeman, and Holzinger in the United States in 1934, 44 in that of Shields in Great Britain in 1962, 12 in that of Juel-Nielson in Denmark in 1965, 53 in that of Burt in Great Britain in 1966.

This last study is based on a reasonably large sample, but its findings are highly questionable. Cyril Burt (1966), who died in 1971 at the age of 88, ruled supreme over British psychologists for almost thirty years. A government adviser on educational matters, his studies on the IQ of twins are, by far, the most often cited. The conclusion of his numerous publications was that in the case of IQ, "broad-sense heritability," representing the proportion of variations due to genetic differences, was in the order of 86 percent. His observations were used extensively by certain psychologists who emphasize the role of genetic inheritance in intellectual activity. In 1974, however, psychologist L. Kamin published an analysis of Burt's work, which revealed strange coincidences: in a study published in 1955 on 21 sets of twins, Burt reported a correlation coefficient of 0.771 for twins reared apart; a study published in 1958 on more than 30 sets again reported 0.771; his final publication in 1966 on 53 sets reported 0.771 yet again. Because of this surprising consistency, C. Burt's methods were subjected to more careful scrutiny. It was found that they in no way corresponded to what is normally expected of scientific work. The tests used are not specified exactly, the sex and age of the children are not indicated in all cases; doubts can even be raised as to the actual existence of some of the twins studied. Moreover, in 1976, the *Sunday Times* medical correspondent reported, after long and meticulous investigation, that the two collaborators who cosigned some of Burt's articles and assisted him in his observations and calculations left no trace on the records of London University, where they were supposed to have worked, and may never have existed. Paleontology had its "Piltdown Affair," psychology its "Burt Affair" (Gillie).

Recently, certain psychologists favorable to Burt announced that they had tracked down one of his collaborators. These incredible ups and downs are of little interest; the only real issue is to decide whether Burt's data can be used in scientific work. According to the researcher who is least suspect of partiality against Burt, the answer is: no. In fact, A. Jensen, with whom the current argument about "intelligence and genetics" originated and who had relied mainly on Burt' conclusions, has stated very clearly that the latter's observations "are of no value for validating the hypotheses" (Jensen 1974; Wade 1976).

Of the other twin studies, only that of James Shields is based on a reasonably large sample. Unlike Burt, this author gives all the necessary details about his observations. Even these have to be used with caution, because the sample he studied is not very representative of the population (twice as many girls as boys, many children from very poor social classes). Furthermore, these children, although "reared apart," often lived together during a good part of their childhood; of 44 pairs, 31 were reared by related families, 4 of which attended the same school. It is difficult to agree that these twins were subject to environmental differences comparable to those of individuals chosen at random.

Lastly, the available data are scanty: 31 pairs in Newman et al. and in Juel-Nielson, 13 pairs actually "reared apart" in Schields. How can one seriously say, based on such slight evidence, that the heritability of IQ has such and such a value? In the previously cited article, L. Kamin (1976) shows that the observations currently available do not allow one to reject the hypothesis of zero heritability.

Another line of research has involved the study of adopted children. By comparing the correlation between the IQs of these children and those of their biological parents on the one hand and those of their adoptive parents on the other, one could hope to reach an estimate of the "broad-sense heritability" for this trait. It is clear that there are numerous difficulties with this kind of study, in particular the fact that neither adopted children nor adoptive parents can be considered as drawn at random from the population.

Based on the few data available, Christopher Jencks (1972), professor of sociology at Harvard, estimated the heritability of IQ at 0.45. More precisely, he estimated that 45 percent of the total variance could be ascribed to the effects of the genes, 35 percent to the effects of environment, and 20 percent to the interaction between genome and environment. Not only did Jencks attempt an estimate of all three terms of the fundamental equation outlined on p. 136, without, like most researchers, ignoring the third, but he gave an estimate of the "confidence interval" for the numbers he proposed. This interval is very broad: ±20 percent. In other words, the data suggest a heritability of between 25 and 65 percent. This is, of course, a high degree of imprecision, but it is a matter of rigor to state it clearly.

Finally, we must admit that we are currently unable to propose a figure for the heritability of IQ; honesty requires that this inability be recognized.

This does not mean that new research, carried out with due regard for scientific protocol, could not gradually lead to estimates with real scientific value. The necessary work will be slow and costly. Before undertaking it, is it not sensible to ask whether it is likely to be useful? Suppose that, in ten or twenty years, we find that IQ has a heritability of 0.43 ± 0.05 among Jurassian peasants, of 0.51 ± 0.06 among the Eskimos of Greenland, 0.70 ± 0.05 among the Bassari of Senegal; what can these results be used for? Only for predicting, within each of the groups, the average IQ of the descendants of couples whose IQ is known. However, in this prediction, the imprecision of the various terms will have to be allowed for: if Mr. X and his wife, Jurassian peasants, both have an IQ of 120 ± 10, their children will have IQ scores dispersed over a range of ± 21 around an average with a confidence interval from 102 to 116. In other words, these children will have 95 chances out of 100 of having an IQ of between 86 and 132, while this range is between 70 and 130 for a child chosen at random from this group.

The imprecision of this result is partly due to the low heritability (0.43 ± 0.05) that we assumed. However, even a high heritability leads to an almost equally vague result. For instance, in a population where the heritability for this trait is 0.70 ± 0.05, the range of dispersion works out to be 95 to 133. Is such a vague conclusion of the slightest interest?

In fact, estimates of the heritability of IQ are not intended to provide this kind of information. In practice, they are used by researchers who are looking at an entirely different problem: analysis of the differences observed between social classes or between races. However, the concept of heritability has no application whatever in this domain.

Inequality of IQ According to Social Class and Race

Jensen's (1969) first published remarks on the genetic determinism of IQ appeared in the *Harvard Educational Review*. They caused a considerable stir only because of the comparisons made between whites and blacks: after numerous tests in both communities, it was found that the average score of blacks was 15 points lower than that of whites. Jensen, relying at the time on C. Burt's conclusions, believed IQ differences to be 80 percent genetic in origin; the intellectual inferiority of blacks, as measured by the tests, therefore reveals an innate biological inferiority, which cannot be remedied.

This article is to be commended for saying bluntly and openly what many people thought privately and for making the underlying argument explicit. Whether its conclusion pleases or scandalizes will depend on one's personal sensitivity; what matters is to establish whether it corresponds to reality.

What we know of the concept of heritability allows us to say that this line of thought is based on a misinterpretation: heritability, which can only be defined and measured within a group, cannot under any circumstances be used for the analysis of differences between groups. American geneticists M. Feldman and R. Lewontin (1977) illustrate this impossibility with a striking example. Among Parisians, whose origins vary widely, darkness or fairness of complexion is a trait greatly influenced by genetic factors and therefore highly "heritable." Compare a group of Parisians who have just come back from winter sports with a group who remained in Paris; can we use this high heritability to claim that the difference between the average complexion of the former, which will be darker, and that of the latter, which will be lighter, is due to a genetic difference? When Jensen and the few psychologists who followed in his footsteps use the heritability of IQ to ascribe the differences in IQ scores between two races to their genetic heritages, they make an equally grave error of logic.

Apart from this misinterpretation, which completely invalidates the argument, many other objections can be raised to Jensen's line of thought. Some are related to what we said on the subject of the instability and imprecision of IQ scores. Others, still more serious, concern the obvious link between the tests used to measure IQ and the characteristics proper to our culture; these tests were developed based on the performance of white children and adults raised in Europe or North America. All the values implicit in their upbringing and formal education have, inevitably, shaped them; how could such tests be of any assistance in judging people immersed in a totally different culture?

In response to this objection, certain psychologists claimed to have developed "culture-free" tests, that is, tests where the result would be independent of the cultural environment of the individual. This is a typically theoretical solution; to be sure, "culture-free" tests would be very interesting, but are they possible? Their objective is to measure intellectual skills and these can only develop within a specific culture; intelligence without culture has scarcely any more meaning than the geneless child to whom we referred earlier.

Finally, the difference found between black and white Americans can be fully explained by the fact that these two groups do not benefit from the same cultural environment, as was well-known before doing any tests or calculations. As for assertions on biological or genetic differences based on this result, they depend on arguments that have no logic whatever.

Similar remarks can be made about the differences found between the various social classes or professional groups. The most ridiculous remarks on this subject appear in a recent book by English psychologist Hans Eysenck (1977). Relying on the results "of a certain number of empirical studies carried out in various countries" (the vagueness of the terminology shows how far this is from being scientific), he draws up an IQ table for the various professions. Here is an extract from it:

140 : Senior executives, professors, scholars and researchers . . .
130 : Middle executives, surgeons, lawyers. . .
100 : Sales people in large stores, train- and truck-drivers . . .
 90 : Gardeners, upholsterers . . . (*Sic!*)

Combined with the dogmatic proclamation that IQ is 80 percent determined by genetic heritage, this kind of table is intended to show that social inequalities are the result of genetic inequalities that cannot be remedied. This is just another manifestation of social Darwinism, which we have already mentioned.

Is it responsible, is it honest, to allow this kind of classification to be published, without taking any of the precautions normally taken by researchers before assigning precise numbers to different categories of people? It is merely ridiculous to persuade professors that their genetic heritages are better than those of lawyers or surgeons. However, it is criminal to persuade "gardeners" or "upholsterers" that their genetic complement puts them at the bottom of the intellectual ladder and that their children will be branded from the moment of conception with this inferiority.

The priggish pedantry is so obvious that one is tempted to shrug one's shoulders and laugh. However, the subject is very serious: in the name of such statements, the entire life of some people may be sacrificed, in particular during that trying obstacle race which schooling has become. The segregation of entire groups and their exploitation can be presented as justified because it is in accordance with the conclusions of science. Scientists, in view of this use or abuse of their work, without regard for the rules that they had imposed upon themselves to insure that the words and figures being manipulated correspond to a precise meaning, have a duty to build a barrier of rigor and to spread their opinion as widely as possible. This is what American geneticists attempted to do in organizing a huge referendum.

In France, the attempt to clarify the situation took another form. The World Movement for Scientific Responsibility devoted its first meeting, in the Sorbonne in March 1977, to reflection on "Genetics and Measuring Intelligence." For two days, psychologists and biologists compared their methods and their concepts, the aim being to avoid misinterpretations in going from one discipline to the other. The concept of heritability was naturally at the center of the discussions. Our remarks here owe a great deal to the work accomplished at this meeting.

The Referendum of the Genetics Society of America

The discussions triggered by Jensen's initial article became so heated that the Genetics Society of America, which includes about 2,600 geneticists from Canada and the United States, tried to establish a limit between what can be considered sceintific conclusions and what are merely arbitrary statements. The first text was published in 1975. It reviewed the various concepts and data relevant to the discussion (IQ, heritability, differences between classes and races . . .) and stated in particular:

1. The limitations to the significance of IQ are especially marked when children from different cultural groups are being compared.

2. Although a major component of IQ variation within a culturally and economically homogeneous group may have a genetic basis, this hypothesis has yet to be tested.

3. There is no convincing proof of a genetic difference in intelligence between races.

4. We feel that geneticists can and must speak out against the use of genetics for the drawing of social and political conclusions from inadequate data.

More than 1,100 responses were returned, 95 percent of which expressed agreement with this text. Fearing no doubt that the nonresponses meant, at least in part, some degree of disagreement, the Society circulated, in January 1976, a second text containing certain nuances compared to the first; on the points outlined above, it stated:

1. The interpretation of IQ scores is especially troublesome when comparisons are made between different cultural groups. These limitations must be borne in mind in any genetic analysis.

2. Although there is substantial agreement that genetic factors are to some extent responsible for differences in IQ within populations, those who have carefully studied the question disagree on the relative magnitudes of genetic and environmental influences, and how they interact.

3. There is no convincing evidence as to whether there is or is not an appreciable genetic difference in intelligence between races.

4. We feel that geneticists can and must speak out against the misuse of genetics for political purposes, and the drawing of social conclusions from inadequate data.

It is obvious that the text was very carefully edited. The nuances compared to the first one are subtle but do not alter the import of the referendum. This time, 1,488 responses were returned, 94 percent of which were in agreement (Russel).

A majority, even as massive as this one, in favor of a certain position, does not necessarily mean that this position is right. To make such a claim is merely another version of the argument from authority, which we have too often condemned. We have stressed this referendum only because we wish to show how far from the truth is the supposition that the "Jensenist" theories are approved by almost all scientists. This is, however, the basic argument of a book (Herbert) published in 1977 which, eight years after Jensen's initial ideas on "Race and Intelligence," tried to spread those ideas in France; incidentally, the authors hid behind a pen name, which proves, if not their courage, at least their lucidity as to their work's lack of value. Accumulating quotations, they try to show that the opponents of the "hereditarist" theses are a tiny minority of "egalitarian militants" subject to the unanimous condemnation of the scientific world. The Genetics Society referendum proves that the exact opposite is true.

Search for Truth or Manipulation of Opinion?

Debating a scientific issue in public can be beneficial; allowing every citizen to reflect on a subject as important as the relationship between genetic heritage and intellectual activity and then to express his considered opinion is an excellent exercise in democracy. However, there is a great danger of the debate becoming personalized and of precise and nuanced statements being replaced by slogans and simplistic catchwords. The way in which some press articles present the problem unfortunately illustrates this danger. Scientists are put into two distinct categories: the "hereditarists," who accept that intelligence is determined mostly by the genetic heritage, the "environmentalists," who claim, on the contrary, that environment plays the greater role. The latter, among whom most geneticists are to be found, are also presented as "egalitarianists" who deny all difference between the biological potential of individuals (de Benoist).

This last label is particularly inaccurate. How could a geneticist, whose leitmotif is diversity, claim that all genetic heritages are "equal"? He has the opportunity to observe the marvelous diversity of these heritages, which are all different; but different is not synonymous with "unequal." Two people can be considered unequal, in the sense that one is superior to the other, only if just one trait is considered; taken as a whole, they can only be different; these two words are far from being equivalent, one implies a hierarchy, the other does not.

The eighteenth-century philosophers did not claim that all humans are equal, which would have meant nothing, but that they are "equal as regards rights," which indicates a political stand. We can say that people today are "unequal in their access to wealth or education," which is a verifiable fact; to say however that they are "equal" or "unequal" in the absolute is completely nonsensical.[2]

These misleading simplifications, which can serve only to manipulate public opinion, must be condemned and the problems formulated clearly.

In conclusion, let us endeavor to single out a few points likely to draw fairly general agreement:

2. This debate is typical of the erroneous interpretation of mathematical words and symbols. The term "equal," represented by the symbol = can be applied:

—— either to two numbers: $x = y$ means that x and y have the same value,
—— or to two sets: $A = B$ means that A and B are made up of the same elements.

The opposite of the term "equal" is not the same in both cases:

—— if two numbers are involved, nonequality means that one is greater than the other, which is written $x > y$ or $y < x$,
—— if two sets are involved, nonequality means that all their elements are not identical; they are different, which is written $A \neq B$.

We must remember these three symbols

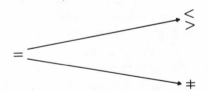

The finding that two objects are not equal implies that one is greater than the other only when these objects are numbers; in all other cases, one can only state that they are different. Any grade twelve student would find this quite obvious. It is remarkable that certain psychologists who observe nonequality conclude that there is necessarily superiority and inferiority.

1. Intellectual activity requires an organ that is constructed based on genetic information, and the training of this organ in the course of a certain human experience that is inadequately described by the word "environment."

2. The ontogenesis of the central nervous system, like that of any organ, is dependent on the genetic heritage: however, this does not mean that this system is genetically defined in all its details. J. -P. Changeux (1977), for instance, stresses the difference in order of magnitude between the number of neuronal connections—10^{14}—and the number of genes—10^5.

3. The intellectual tool that we have at any particular time is the result of the genetic information that we have received, of the material at our disposal for its development, and of *the use that we have made of it:* this last essential point is often neglected.

4. Any two individuals have, inevitably, different genetic heritages (monozygotic twins excepted) and have had different life experiences. The intellectual tools available to them are, of course, different. This difference may be expressed as a difference in the particular aspect of intelligence that IQ is intended to measure. However, in all but the exceptional case of monozygotic twins, we have no means of ascribing this difference to one cause rather than another (the concept of heritability, in particular, cannot be used for this purpose).

5. Two groups of individuals (for instance, two "races," if we consider ourselves capable of defining them) have, as a whole, the same genes, but with different frequencies. The genetic information for cerebral ontogenesis may therefore be distributed differently from one population to another. Simultaneously, the life-styles, the cultures which mold the intellectual faculties, are usually very different. It is not in the least surprising that a difference in "average IQ" is found. However, we have no way of ascribing this difference to the different causes (without exception this time, since there are no twin populations).

6. Even if the difference found between the IQ of whites and blacks in the United States corresponded to an objective measure, it is completely illogical to conclude from it that the "average" genetic heritage of blacks is "unfavorable." Even if we were to show

that individual differences within each community are almost totally due to genetic differences, we would have no right whatever
to impute the difference found between the averages of the two
communities to a genetic difference. In general, the discovery that
individual differences within a population are due to a specific cause
does not allow us to ascribe differences between populations to this
same cause.

7. *Any attempt at justifying social inequalities that is based on measures such as IQ and concepts such as heritability constitutes therefore
a fraudulent use of science. Any program claiming to improve the "intellectual potential" of a group through eugenic planning can only be a
moral swindle.*

The fact that research leads to a "measure" does not necessarily
mean that it is scientific, opportune, or even simply harmless. By
way of analogy, Noam Chomsky (quoted in Morin and Piattelli—
Palmarini, eds. (1974:803)) imagines a situation in which psychologists or geneticists proposed to the German authorities, in the
1930s, to study "potential for the acquisition of wealth" in the
Jewish population, and to estimate to what extent this trait is genetically determined. These researchers could have claimed that they
were merely contributing to the advancement of knowledge; they
could have accused any opponents of being afraid of the truth; their
enterprise would, nonetheless, have had but one end, justification
of genocide. Without looking for imaginary examples, it is sufficient to recall certain kinds of research done by Nazi doctors in
concentration camps to realize that the advancement of knowledge
is not a justification for everything.

It is evident that the reflections on the "innate and the acquired," determinism and freedom, have gone completely off track
and into an impasse in turning to research on the heritability of
IQ. This concept, developed and used by geneticists interested in
improving certain well-defined quantitative traits in plants and animals, cannot be used in studies concerning human beings. It is
because of this fact of logic that certain American geneticists have

proposed to cease research in this direction. Contrary to what certain psychologists have claimed, this is not a sign of deliberate obscurantism, or of a refusal to know the truth. Rather, it is simply the logical consequence of a precise analysis of the objects and concepts being manipulated. Discussions about the sex of angels occupied excellent minds, to no avail, for quite a long time. Is not research on the heritability of IQ just as futile?

An accurate answer to a badly or incompletely formulated question can actually be deceptive. We now know that science cannot be neutral; its principal aim must not be to reply to questions, but to clarify the exact meaning of these questions.

EIGHT

The Temptation To Act

THE CURRENT population explosion makes it obvious that intervention and planning are necessary. There is an urgent need to restrict population size; is it not natural to treat the qualitative and the quantitative simultaneously, to endeavor simultaneously to "improve" the human race?

The Population Explosion

The demographic history of humanity can be summarized, without oversimplification, by distinguishing three periods separated by two "revolutions."

In the prehistoric period, the total number of people on the entire surface of the earth was somewhere between a hundred thousand and a million. According to J. N. Biraben (1977), from whom the following figures are taken, this figure grew from under 800,000 around the year 35,000 B.C. to 4 or 8 million around the year 10,000 B.C. The slowness of this progression and the smallness of this number are explained by the limited resources available to hunter-gatherer populations.

The first "demographic revolution" was brought about by the invention of agriculture. Quickly, the population grew, from 7 or 8 million to 80 million by 5000 B.C. The increase slowed subsequently: 250 million at the beginning of our era, 800 million at the end of the eighteenth century.

This is when the second demographic revolution occurred. It was due to the success finally achieved in the struggle against sickness and death, particularly among children; thanks to more efficient medical care and especially to more sensible hygiene, life expectancy increased (for instance, in France, it increased from 38 years for women at the end of the eighteenth century, to 76 years, exactly double, at present; the proportion of children to die before the age of 15 decreased simultaneously from 34 percent to 3 percent). The result is that the curve of human population size has taken on a completely different shape: a slow and noticeably linear growth pattern has given way to an exponential increase that is truly explosive: 1.2 billion in 1850, 1.6 billion in 1900, 2.5 billion in 1950, 3.6 billion in 1970, more than 4 billion today, 6 billion before the end of this century. At the present rate, the human population increases in just three years by as many people as were alive at the time of Jesus Christ.

This explosion is illustrated by the following comparison: since the emergence of Homo sapiens—the estimated date for this event is of little consequence—the number of years lived by humanity as a whole is in the order of two thousand billion; everyone alive today, our 4 billion contemporaries, has lived or will live between them all about two hundred billion years; in other words, the present human population will accumulate one tenth of the total time lived by humanity as a whole since its origin.

This exponential development, which may seem to be just one phenomenon among many, since its description in numerical terms tends to camouflage the reality, can only lead to catastrophe if collective action is not taken quickly.

Certainly, there can be considerable differences of opinion as to the maximum number of people that the earth can support and nourish. Serious studies, but naturally with widely divergent results, have been carried out based on food, energy, or water supplies. In order to estimate the final cutting off point for this expansion, we will limit ourselves to the resource which least lends itself to discussion: space. The total land surface not under water is about 150 million km^2. Even if it were somehow possible to make deserts, mountains, and the frozen territories of Greenland and the Antarctic habitable in such a way as to extend Japan's current population density (300 people per km^2) to the entire earth, this would still not allow for more than 45 billion people. At the present growth rate, 1.9 percent per annum, this limit will be reached in the year 2100, that is, in five generations.

Suppose that, as in certain overpopulated regions of Asia, it then becomes possible to organize life on water and that the density of 300 people/km^2 becomes extended to the planet as a whole, including the oceans; this would allow for a population of 150 billion. This figure may seem considerable, providing some margin for expansion; in reality, again at the current growth rate, it will be reached towards the end of the twenty-second century, in two hundred years.

Arguments about maximum population size are therefore quite futile. Even if they are based on differences of up to several tens of billions of individuals, they mean, in terms of time, only a few decades: one estimate of maximum size that is double that of another corresponds, at the current growth rate, to a difference of only thirty-five years, barely more than a generation, in the date at which this maximum will be reached.

Whether chosen or imposed, the third demographic revolution, that is, the switch from exponential growth to quasi-stability, cannot be avoided. Either we abandon medical intervention, which is so successfully fighting disease, or we limit the number of births. Few people find the first option acceptable and, therefore, the consequences of the second must be seriously contemplated.

In many Western countries, this revolution has, of course, already occurred, which prevents their inhabitants from being fully aware of the global problem. In some, stability has even been surpassed and a decrease has begun. However, these are merely local phenomena, with little influence on developments in the rest of the world: there remains less than a century in which to extend the new regime to humanity as a whole and to reduce the overall growth rate to zero.

A change of this kind will obviously have major repercussions on social structures and on individual attitudes. It will change the way in which culture and genetic heritages are transmitted. We will confine ourselves here to a very specific and seldom mentioned aspect of the change: the consequences of the discrepancies in the rates at which the new regime will be achieved, and especially the variance of growth rates once stability is reached.

The Consequences of a New Demographic Regime

Accepting zero growth rate means accepting a culture where the right to procreate is subject either to extremely tight regulation (as in certain Indian states since quite recently) or to a type of social pressure that in practice eliminates deviants from "normal" behavior (as in contemporary Western societies). A culture of this kind has traits depending directly on the age structure of the population. The famous age "pyramid" gradually becomes (to use another image from ancient Egypt) an "obelisk." The successive age categories, which hitherto got progressively smaller, are henceforth almost constant in size, right up to the very oldest categories (over 80, 90, or 100 years old), which diminish very rapidly.

These two types of population structure are to be found at the present time in two actual cases, the "pyramid" model in Mexico and Egypt on the one hand, and the "obelisk" model in West Germany and Hungary on the other. The distribution of ages in these coutries is as follows:

	0–14 yrs	15–64 yrs	+ 65 yrs	
Mexico	46.2 %	50.1 %	3.7%	"Pyramid"
Egypt	42.4 %	54.4 %	3.2 %	
West Germany	23.2 %	63.6 %	13.2 %	"Obelisk"
Hungary	20.2 %	67.4 %	12.4 %	

Given these numbers, it is easy to imagine how the internal dynamics of these societies must differ; independently of differences in political regime or ideology, the importance of the older people, the pressure exerted by the young, the load to be borne by the adults will be quite different. The institutions developing within societies subject to such different numerical constraints will, inevitably, be very different.

A time-lag between the dates at which the various societies begin the "third demographic revolution" therefore leads to discrepancies and tensions, the consequences of which are difficult to predict. Cultural differences, which are at the origin of these time-lags, can only be further accentuated by the consequences of these different patterns of demographic change. We are confronted with a process which does not regulate itself automatically, unlike most of those found both in the living and in the inanimate worlds. This is, therefore, an explosive phenomenon which our minds are ill-prepared to assimilate fully. We are accustomed to self-regulating phenomena, the reason being that non-self-regulating phenomena cannot persist; therefore we can have but very limited experience of them. However, it would be absurd to conclude that they do not exist and need not be envisaged.

The time-lag in reaching the required demographic stability has obvious consequences for the relationships between the sizes of the various human groups. The current global growth rate, 1.9 percent per annum, is an average beneath which major differences between populations are hidden: while there are some populations which are still increasing at a rate of over 4 percent per annum, there are some Western European populations that have already reached stability.

When Mexico emerged, round 1920, from a long period of revolutions and massacres, its population was the same size as it had been four centuries previously on the arrival of the Spanish conquerors, 14 million, barely more than a third of the population of France. A half-century later, in 1970, Mexico and France had populations of the same size, 50 million. However, the annual rate of increase in Mexico is 3.4 percent with a tendency to accelerate (in 1950 it was only 2.7 percent, in 1930 1.1 percent), while in France it is less than 0.5 percent and is tending towards zero. In the event of these trends continuing until the end of the century, Mexico's population would be greater than 130 million in the year

2000, while that of France would not have reached 60 million: a time-lag of one generation will have been sufficient for the relationship between the two population sizes to go from 1 to more than 2.

These reversals do, of course, affect the balance of power between groups. This does not depend solely on numerical factors, but the pressure of increased numbers may trigger defense reactions on the part of those who sense that this component of their power is gradually escaping them.

The changes to come will inevitably lead either to a major shift in cultural ascendancies or to the increased use of a kind of pressure that is independent of numbers and therefore nondemocratic by definition.

One thing is certain, the world at the end of this century will be different from what it is now, both in its actual state and in the speed with which it will be changing. Our mind is accustomed to comfortable extrapolations, but an explosion cannot be extrapolated.

My children were born in a world of 2.6 billion people; when they reach my age, they will be surrounded by 6 billion people. This human deluge, which is about to submerge the earth, seems to lend support to arguments in favor of a policy of selection, of quality: it is suggested that the best should be chosen from these vast masses and given the means of complete fulfillment; as for the others, is not the blissfully dumb fate of the "Epsilons" imagined by Aldous Huxley the best that they can hope for?

Past Recourse and Present Allusions to Eugenics

The wish to influence the "quality" of humanity is apparent in the attitude of numerous people. One extreme case is represented by the small village of Tenganan in the island of Bali, studied by G. Breguet (thesis in preparation); because of its religious option, it has been totally isolated genetically from the neighboring villages since the fourteenth century. The god Indra, mythic founder of the community, requires that those who serve him during ceremonies have a perfect body; people with "defects" (blindness, harelip, leprosy, etc.) are not allowed to have children. This is just a village of 300 people. Eugenic measures affecting much larger communities are present in everybody's mind.

National Socialist Germany is, of course, the country that showed the most determination in dealing with the problem. From 1933 onward, legislative texts deal with the sterilization of certain individuals, the regulation of marriages, the putting aside of certain ethnic groups. "The improvement of the race" was assured both through the elimination of defective individuals and through the organization of systematic "razzias," in Poland, for example, during the occupation; hundreds of thousands of little girls, with traits that were considered favorable, were sent to German families, where they were to be reared until puberty, before being fecundated by young SS members. Everything was planned, each girl was to supply three children after which they were to be eliminated (Hillel). The units responsible, at all levels, were advised by geneticists. It was a geneticist, whose scientific value is not in question, Otmar von Verschuer (1943), director of the Anthropology, Human Heredity and Eugenics Institute in Berlin, who, in a widely documented book translated into French, wrote enthusiastically that: "The head of the German ethnoempire is the first statesman to have made the facts of hereditary biology a guiding principle for the management of the State." Based on genetic considerations, he stated: "The politics of the present require a new and total solution to the Jewish problem" and declared (in 1943, note well this date and all that it implies): "The Tzigane question will soon be settled." This is an example of what scientists are capable of writing and promulgating. There are other examples, doubtless less dramatic, but which denote a very similar state of mind.

In a country like the United States, where there is intense immigration, the idea that social problems resulted from the poor quality of the new immigrants gained easy acceptance. The discovery of the "laws of heredity" made it possible to give a seemingly biological justification to this feeling. In a remarkable study, Jon Beckwith (1976), of the Harvard Medical School, gathered numerous quotations indicative of this state of mind. For instance, the geneticist Davenport stated: "Social reform is futile and the only way to secure innate capacity is by breeding it," while Professor McDougall, chairman of the department of psychology at Harvard, called for "the replacement of democracy by a caste system based upon biological capacity with legal restrictions upon breeding by the lower castes and upon intermarriage between the castes."

These statements led, in certain cases, to concrete measures, especially the sterilization of individuals with defects considered to be hereditary. In his very detailed review of this problem, J. Sutter (1950) points out that the legislation of some states was aimed at "sexual perverts," of others at "habitual criminals" or at syphilitics. Between 1907 and 1949, 50,000 sterilizations were carried out in 33 states, almost half of which were done on the "feebleminded." Other laws forbade marriage, mainly between blacks and whites, but sometimes between Orientals and whites also. It was not until 1967 that these laws were declared unconstitutional.

However, the most important decision concerned immigration. A national committee was set up to study the risk of deterioration of the country's genetic heritage due to the influx of individuals from inferior populations. Professor Brigham, a psychologist and adviser to this committee, notes in an official report: "The decline in intelligence is caused by the immigration of Blacks and of people of the Alpine and Mediterranean races." He demands that "immigration be not only restrictive, but highly selective," and advocates "measures dictated by science and not by politics." These measures were taken: the famous Immigration Act of 1924 severely limited immigration from southern and eastern European countries.

To Brigham's credit it must be added that, in 1930, he referred to his study as "one of the most pretentious of those comparative racial studies" and says that it was "without basis." However, the Immigration Act was not, for all that, revoked immediately; not until 1962 did Congress finally change it.

Few political or scientific leaders would now risk openly advocating eugenic measures, but one often hears statements aimed at preparing people's minds to accept them, for instance, this sentence from a top-level education policy maker in France, published in an evening newspaper: "The *genetic* potential for success is greater, statistically, in the descendants of individuals who have themselves reached a higher than average level of achievement" (Capelle).

One can see how easy it is to go from a statement of this sort to justifying a "brave new world" Huxley-style. As soon as one accepts that the role played in society and the services rendered to it are directly related to the genetic complement of each individual, it is natural to think of orienting human reproduction. We are actually heading toward a society where the right to reproduction will, inevitably, be limited. In these circumstances, it is almost inevitable to end up thinking like Bentley Glass (quoted in Beckwith): "The right that must become paramount is not the right to procreate, but rather the right of every child to be born with a sound physical and mental constitution, based on a sound genotype."

It is the last part of this sentence that poses a problem. We can agree on the right of children to be born with a "sound phenotype"; however, we have shown throughout this book that the transition to a "sound genotype" is far from being as simple as Glass assumes it to be. How is the quality of a genotype to be judged?

The Difficulty in Judging

We are accustomed to judging phenotypes. A particular trait can validly be described as good or bad; even if there are divergent views, a certain consensus often appears. To be sure, this judgment is a function of a certain milieu and of a certain objective: the mentality of a kamikaze is, in general, a disadvantage to the individual, but it may be highly advantageous to society if this attitude enables it to defend itself better. We are accustomed to reversing our judgments when the circumstances or objectives change.

This kind of flexibility cannot be valid at the genotypic level, because a gene lasts infinitely longer than the individual who carries it. We have already mentioned this difficulty in relation to diabetes: some of the gene combinations responsible for this disease are no doubt beneficial in time of famine. How is one to deliver judgment on these genotypes which will continue to be transmitted for millenia and will be alternatively harmful or beneficial?

This also applies to the famous *S* gene for abnormal hemoglobin; while it causes the death of homozygotes, it protects heterozygotes from malaria. If one's aim is the survival of the group in a malarial region, this gene, though fatal to some, is highly favorable; without it, the group would become extinct.

When dealing in chapter 5 with the theories of evolution, we mentioned arguments that are presented as non-Darwinian. These arguments try to explain the evolution of the living world with the least possible use of the difficult concept of "selective value." They do not deny the existence of selective pressures, but they do remounce the attempt at unraveling their intricacy; they proceed "as though" reality, at each instant, were the result of a lottery, like a ball chosen at random from an urn overflowing with an infinity of possibilities. This attitude implies an admission of impotence in the face of the judgment to be made with regard to the selective value of each gene. By renouncing judgments of this kind, scientists can develop a theory which, though it may fail to explain details, is consistent with the available data. The "neutral gene" hypothesis is not only a mathematical contrast to the hypothesis of "selected genes," it implies a particular conception of the "realm of genotypes." Unidimensional, black and white judgments, based on two distinct points of reference indentified as "good" and "bad," have scarcely any place in this realm. The assumption of the "neutrality" of the genes corresponds to a refusal to work with a simplistic model that is a mere caricature of the richness of the real world.

How could this judgment, which, except in extreme cases, cannot be made with regard to individual genes, conceivably be made with regard to collections of genes, or the collective genetic heritage of a population? The criterion is no longer the future of one gene or, in the short term, the future of one individual, but the future of a human group as a whole, its capacity for renewal and for struggle, from generation to generation, against the erosion of time. This struggle is all the more efficacious as the possibilities for transformation, for adaptation to a changing environment, are greater, therefore, as the collection of genes is more varied. Even if we give up on the attempt at distinguishing "good" and "bad" genes, we can nevertheless specify what constitutes a "good" collective genetic heritage: it must be diverse. What matters is not the average value of the genes present in this heritage, but the diversity of these values.

Many of the arguments developed from the "fundamental theorem of natural selection," which we mentioned when describing theories of evolution, are based on confusion between the average of the selective values of individuals who compose a population and the selective value of the population itself. This confusion is of the same order as that stressed by mathematicians who point out the difference between an "element" of a set and a "part" of a set, between the notion of belonging and the notion of inclusion. The selective value of an individual refers to his capacity for successfully competing against the other individuals in the same group; the selective value of a population refers to its capacity for successfully competing against the other groups, of the same species or of other species, within a certain ecological community. The two concepts are defined within two distinct frames of reference. Both are a function of the frequencies of the diverse genes, but these functions may be quite different: it is easy to imagine situations where, with all individuals approaching the maximum individual selective value, the group becomes homogeneous and loses all its capacity for structuring and organizing itself to the best advantage.

This distinction is important to the understanding of evolution. The famous "struggle for survival," which, according to Darwin, is its driving force, operates at two levels: competition between individuals within a population and competition between populations in an environment with limited resources. This distinction is essential to any arguments aimed at founding eugenics.

In this perspective, the genetic value of an individual for the collectivity is not a function of the actual quality of his genes, but of the fact that these genes are uncommon. The goal is, therefore, no longer the "improvement of individuals" but the preservation of diversity. The goal of conscious and responsible management of the genetic heritage is, therefore, quite different from that pursued by those in favor of eugenics: it is no longer a matter of eliminating bad genes and favoring good ones, but of safeguarding the genetic richness that comes from the presence of diverse genes. We are a long way from the simplistic position of those who propose various measures (prohibition of certain types of marriages, sterilizations, etc.) with a view to promoting the genetic quality of a population by increasing the number of "good" genes.

Eugenics is, no doubt, an extreme example of the perverse use of science: it is in the name of science that some of the worst horrors have been suggested and sometimes perpetrated. These abuses are leading many of our contemporaries to question the validity of the scientific enterprise: what began as a work of liberation has become suspect, so great is the risk that it will lead to the domination of the majority by a minority. Is the advancement of knowledge, long synonymous with the advancement of humanity, not going to lead to the subjugation, if not to the annihilation, of our species? This widely felt anguish explains the rapid success of the World Movement for Scientific Responsibility, founded three years ago by Robert Mallet. We saw that this responsibility was first exercised on the question of "genetics and intelligence measurement," but many other issues require similar attention.

We have just seen that genetic richness comes from diversity. It seems clear that this observation extends beyond the domain of biology: the richness of a group is made up of "its rebels and its mutants," in the words of Edgard Morin (1973). We need to understand that others are precious to us insofar as they are different from us. This is not a moral precept that is either freely chosen or imposed by a revealed religion, it is a lesson that genetics teaches us directly.

Is this preaching tolerance? What an ugly word! We recall the unpleasant response given by Paul Claudel, who was being criticized for his intolerance: "As for tolerance, there are establishments specializing in that."[1] To tolerate is to comply half-heart-

---

1. Translator's note: in French, a bawdyhouse is known as "une maison de tolérance," i.e., "a house of tolerance."

edly, it is to be prepared to "put up" with someone, it is, in the negative, to not forbid. It implies a power relationship where the dominant one consents, condescends not to use his power. A person who is being tolerant feels that he is being very good to do so, while the person being tolerated feels doubly despised, both for what he represents or professes and for his inability to impose it. Intolerance, a defence mechanism of the weak or the foolish, is, of course, a sign of infantilism, but tolerance, a concession granted by someone secure in a position of power, is only the first step toward full recognition of the other person; many more are required before "love of differences" can be achieved.

Love of Differences

"If I am different from you, I enrich rather than diminish you," Saint-Exupéry, *Letter to a Hostage*. This obvious fact goes counter to all our reflexes. Our superficial need for intellectual facility leads us to think of and judge everything in relation to a type; but differences are enriching.

Much deeper and more fundamental is the need to be unique, so as to truly "be." We are obsessed with the desire to be recognized as original and irreplaceable individuals. This is what we are in reality, but we are never quite satisfied that our entourage is sufficiently aware of it. What finer gift can another person give us than to reinforce our feeling of uniqueness and originality by being different from us? Instead of attempting to smooth over conflicts and blot out confrontations, we should realize that these conflicts and confrontations must and can be of benefit to us all.

This is possible provided that the aim is neither the destruction of the other person nor the establishment of a hierarchy, but the gradual construction of each individual. The clash, even though violent, is beneficial; it allows each one to reveal himself in his singularity. Competition, which is, on the contrary, almost always insidious, is destructive; it can only lead to each one being locked into an imposed order, a necessarily artificial and arbitrary hierarchy.

The first lesson of genetics is that individuals are all different and cannot be classified, evaluated, or placed on a hierarchy: the definition of "races," which is useful in some kinds of research, cannot fail to be arbitrary and imprecise. The search for the "better" and the "best" is futile. The one quality that is considered proper to humanity, intelligence, and which is the source of such pride, essentially escapes our analytic techniques. Past attempts at biological "improvement" of humanity have sometimes been simply ridiculous, usually criminal with regard to individuals, and devastating for the group.

Luckily, nature is marvelously robust in the face of humanity's misdeeds: the genetic flux continues its work of differentiation and maintenance of diversity, almost unaffected by human actions. The "realm of phenotypes," within which we live, fortunately has very little chance of affecting the "realm of genotypes," on which our future depends.[2] The transformation of our genetic heritage is a temptation, but this kind of intervention will, it is to be hoped, be beyond us for a long time to come.

This reflection can be transposed from genetics to culture: the cultures that humanity has developed are wonderully diverse, and it is this diversity that constitutes the richness of each of us. Because of difficulties in communication, this cultural heterogeneity lasted for a very long time. However, it is now clear that it is in danger of disappearing rapidly. Western culture, as it is commonly called, has made amazing strides toward the aim which it had set itself: material well-being. This success is giving it an unprecedented power of diffusion, which is gradually leading to the

---

2. The only case in which man can hope to soon be able to really influence the genes which he transmits is the choice of the sex of children. According to whether the sperm that fertilizes the egg is the carrier of a Y chromosome or an X chromosome, the child is a boy or a girl. It is very probable that, before many years have passed, it will be possible to separate these two categories of sperm and, through artificial insemination, to conceive a child of the desired sex. Many studies have tried to predict the consequences of this "progress" for our society. Some of them predict a considerable numerical imbalance between the two sexes, with apocalyptic consequences. Others, based on the expressed intentions of the future parents at the time of the investigation, assume that a balance will quickly be reestablished and that the advantages of this technique (especially for limiting the number of births) will supersede the inconveniences (Bodmer and Cavalli-Sforza 1976:673). Once again, we have to admit our total inability to predict the long-term consequences of a possible technical achievement.

destruction of all the others. Such was the fate, to cite but one example among so many others, of the Eskimos of Ammassalik, on the coast of Greenland, whose cultural death under the pressure of "obligatory civilization" has been described by R. Gessain (1970).

When the quality of human relationships and of social harmony in certain groups that we call "primitive" is taken into consideration, the wisdom of imitating our culture seems highly questionable, and one wonders if it will not lead to catastrophe. The price paid for the improvement in the standard of living is terribly high, if harmony and cooperation are replaced by our internal contradictions, our tensions, and our conflicts. Is there still time to avoid the leveling of cultures? Is not the richness of diversity worth the abandonment of certain objectives that are measured in terms of gross national product or even in life expectancy?

This is a very serious question and it must not be taken lightly. It is difficult, when reflecting on it, to remain consistent with oneself, according to whether one asks it in the cosy tranquility of one's library or while briefly sharing the life of one of those groups, some aspects of which we admire, but where children die for want of food and medical attention.

Will we be able to safeguard cultural diversity without paying an exorbitant price for it? Whether wished for or merely endured, a change in the organization of our planet cannot be avoided. This is therefore the time for "utopists" to speak. Some of them present the problem in unexpected terms, for instance, Yona Friedman (1974 and 1975), who wrote a book entitled *How to Live with Others Without Being Either Leader or Slave.* Even when the world they envisage seems too "different" from our own, we can be almost certain that reality will be even more so.

As for this exercise in imagination, it seems that the much condemned generation that is preparing to succeed us has already undertaken it to a large extent. It is our children who must teach us to break out of the daily round, the straitjacket of sickly sweet comfort, the insipidness of regimented daily living, and the insidious death of compliance. But will they know how to build a world where Man will no longer be at the mercy of Man?

The Effect of Curing Diseases on the Evolution of Gene Frequencies

1. Diseases due to recessive genes (that is, genes which are expressed only when homozygous)

Let p be the frequency of the gene d responsible for the disease, $1-p$ the frequency of the normal gene N. We assume that the homozygotes (dd) have a chance of survival to reproductive age equal to k times that of unaffected people ($k=0$ if the disease is always fatal, k is close to 1 if the disease is benign).

At birth, the proportions of the various genotypes are given by the Hardy-Weinberg rule:

Genotype	NN	Nd	dd
Frequency	$(1-p)^2$	$2p(1-p)$	p^2

After selection due to the disease, a fraction $(1-k)$ of the homozygotes (dd) disappears and the proportions become:

$$\frac{(1-p)^2}{1-(1-k)\,p^2} \qquad \frac{2p\,(1-p)}{1-(1-k)\,p^2} \qquad \frac{kp^2}{1-(1-k)\,p^2}$$

and the frequency of the gene d becomes:

$$p' = \frac{p(1-p)+kp^2}{1-(1-k)\,p^2}$$

or, if p is small, approximately:

$$p' \simeq p-(1-k)\,p^2.$$

The change in frequency due to the disease is therefore, between two generations:

$$\Delta p = -(1-k)\,p^2$$

When the frequency of the disease in a population seems constant, the decrease in p due to the death of some of those affected is compensated for by other factors, mutations or the selective advantage of heterozygotes, for instance. Curing the disease will not change these other factors; the equilibrium will therefore be replaced by a progressive increase in the frequency of the d gene, an increase which, for simplicity, we assume to be equal and of the opposite sign to the Δp which we have just calculated.

Let us adopt the generation as the unit of time; equation (1) with the sign changed can be written as: $\Delta p = (1-k)\,p^2\,\Delta t$, or, in differential form:

$$dt = \frac{dp}{(1-k)\,p^2};$$

by integration between the initial generation and the tth generation, one obtains:

$$t = \frac{1}{1-k}\left(\frac{1}{P_0}-\frac{1}{P_t}\right),$$

a relation (equation) which allows us to calculate the time necessary for the frequency of the gene to undergo a given change. For instance, the time required for it to double is:

$$t\ (\text{doubling}) = \frac{1}{2\,(1-k)\,P_0}$$

For phenylketonuria, which used to be fatal in the past, $k=0$ and $P_0 = \frac{1}{105}$, hence $t\ (\text{doubling}) = 52.5$.

For cystic fibrosis: $P_0 = \dfrac{2}{100}$, and $t = 25$.

2. Diseases due to sex-linked genes (that is, genes carried by the X chromosome)

Human females have two X chromosomes. Since the gene d responsible for the disease is assumed to be recessive, only homozygous (dd) females will be affected. Males, on the contrary, have only one X chromosome: they are, therefore, affected if this chromosome carries a d gene.

Let p be the frequency of d, $1-p$ that of the normal gene N. At birth, the proportions of the various genotypes are the following:

$$
\begin{array}{ccccc}
& \text{Females} & & & \text{Males} \\
NN & Nd & dd & N- & d- \\
(1-p)^2 & 2p\,(1-p) & p^2 & 1-p & p
\end{array}
$$

Assume that the chance of survival of those affected by the disease is equal to k times that of unaffected people (where k is somewhere between 0 and 1). After selection due to disease, the proportions become:

$$
\frac{(1-p)^2}{1-(1-k)p^2} \quad \frac{2p(1-p)}{1-(1-k)p^2} \quad \frac{kp^2}{1-(1-k)p^2} \quad \frac{1-p}{1-(1-k)p} \quad \frac{kp}{1-(1-k)p}
$$

Females carry $\tfrac{2}{3}$ of the d genes and males $\tfrac{1}{3}$. Their new frequency is therefore:

$$
p' = \frac{2}{3}\left(\frac{p(1-p)+kp^2}{1-(1-k)\,p^2} \right) + \frac{1}{3}\left(\frac{kp}{1-(1-k)\,p} \right)
$$

or approximately, if p is small:

$$
p' \simeq \frac{2p+kp}{3}
$$

which corresponds to a variation in frequency:

$$
\Delta p = \frac{(k-1)\,p}{3}
$$

The curing of the disease, breaking an equilibrium, causes an increase in frequency equal to $\dfrac{(1-k)}{3}\,p$.

By integration, as in the preceeding case, one obtains:

$$t = \frac{3}{1-k} Lt \frac{P_t}{P_0}$$

The time required for doubling is therefore:

$$t \text{ (doubling)} = \frac{2.1}{1-k}$$

If we assume that, for hemophilia, $k = 0.5$, the time required for doubling its frequency is $\frac{2.1}{1-0.5} \simeq 4$ generations.

REFERENCES

Achard, P. et al. 1977. *Discours biologique et ordre social*. Paris: Ed. du Seuil.

Beckwith, J. 1976. *Social and Political Uses of Genetics in the United States: past and Present*. Annales New York Academy, 265:46–58.

Benoist, A. de 1977. "Hérédité de l'intelligence: le débat est ouvert." *Le Figaro*, November 19–20, p. 26.

Bergues, H. 1960. *La Prévention des naissances dans la famile*. Paris: PUF-INED.

Bernard, J. and J. Ruffié. 1966. *Hématologie géographique*, vol. 1, *Écologie humaine, caractères héréditaires du sang*. Paris: Masson.

Biraben, J. N. 1979. "Essai sur l'évolution du nombre des hommes." *Population*, 34:16–26.

Blum, H. F. 1967. "Does the Melanin Pigment of Human Skin Have an Adaptative Value?" In N. Korn and F. Thompson, eds., *Human Evolution*, pp. 362–384. New York: Holt, Rinehart and Winston.

Bocquet, C. "Sélection." *Encyclopaedia Universalis*, 14:849–851.

Bodmer, W. and L. L. Cavalli-Sforza. 1976. *Genetics, Evolution, and Man*. San Francisco: Freeman.

Bois, E. et al. 1978. "Cluster of Cystic Fibrosis in a Limited Area of Britanny (France)." *Clinical Genetics*.

Breguet, G. "Le Village de Tenganan, Bali." Ph.D. thesis, in preparation, Université de Genève.

Burt, C. 1966. "The Genetic Determination of Differences in Intelligence: A Study of Monozygotic Twins Reared Together and Apart." *Brit. J. Psychol.*, 57:137–153.

Capelle, J. 1977. "Les CES ont-ils échoué?" *Le Monde*, February 2.

Changeux, J.-P. 1977. "Déterminisme génétique et modulation épigénétique des réseaux de neurones." In A. Lichnerowicz and F. Perroux, *L'Idée de régulation dans les sciences.* Paris: Maloine-Doin. 1977.

Chapman, A. and A. Jacquard. 1971. "Un Isolat d'Amérique centrale: les Indiens jicaques du Honduras." *Génétique et Populations,* pp. 163–185. Paris: PUF-INED.

Chaventré, A. 1974. "Les Touareg Kel Kummer." In A. Jacquard, *Génétique et populations humaines.* Paris: PUF.

Chaventré, A. and A. Jacquard. 1974. "Un 'Isolat' du Sud Sahara: les Kel Kummer, VII. Conclusions provisoires." *Population,* 29:528–534.

Courgeau. D. 1973. "Les Enfants nés à l'étranger." *Enquête nationale sur le niveau intellectuel des enfants d'âge scolaire.* Paris: PUF-INED.

Crozat. 1827. *Géographie universelle.* Paris: Amable Costes.

Dague, P. 1977. "La Mesure de l'intelligence." *Actes du Colloque "Génétique et mesure de l'intelligence."* Paris: MURS. in press.

Dausset, J. and J. Colombani. 1973. *Histocompatibility Testing, 1972.* Copenhagen: Munksgaard.

Dubertret, L. 1975. *L'Homme et son programme.* Paris: Denoël.

Eysenck, H. 1977. *L'Inégalité de l'homme.* Paris: Copernic.

Feingold J. et al. 1974. "Fréquence de la fibrose kystique du pancréas en France." *Annales de Génétiques,* 17:257–259.

Feldman, M. and R. Lewontin. 1982. "L'Héritabilité au rancart." In *La Transmission,* pp. 205–220. Paris: Fayard.

Fisher, R. A. 1930. *The Genetical Theory of Natural Selection.* Oxford: Clarendon.

Franklin, I. and M. Feldman. 1977. "Two Loci with Two Alleles: Linkage Equilibrium and Linkage Disequilibrium Can Be Simultaneously Stable." *Theor. Popul. Biol.,* 12:95–113.

Frezal, J., J. Feingold, and H. Tuchmann-Duplessis. 1973. *Génétique, Maladies du métabolisme et embryopathie.* Paris: Flammarion.

Friedman, Y. 1974. *Comment vivre entre les autres sans être chef et sans être esclave.* Paris: J.J. Pauvert.

Friedman, Y. 1975. *L'Utopie réalisable.* Paris: 10–18.

Genevois, L. 1975. "Les Nouveaux Blés et la 'révolution verte.' " *Journ. d'Agriculture et Botanique Appliquée,* 22(1,2,3):47–55.

Georges, A. and A. Jacquard. 1968. "Effets de la consanguinité sur la mortalité infantile." *Population,* 23:1055–1064.

Gessain, R. 1970. *Ammassalik ou la civilisation obligatoire.* Paris: Flammarion.

Gillie, O. 1976. "Crucial Data Was Faked by Eminent Psychologist." *The Sunday Times,* October 24, pp. 1–2.

Grouchy, J. De 1973. *Les Nouveaux Pygmalions*. Paris: Gauthier-Villars.

Hartl, D. 1977. *Our Uncertain Heritage: Genetics and Human Diversity*. Philadelphia: Lippincott.

Herbert, J. P. 1977. *Race et intelligence*. Paris: Copernic.

Hiernaux, J. 1969. *Égalité ou inégalité des races?* Paris: Hachette.

Hillel, M. 1975. *Au nom de la race*. Paris: Fayard.

Jacquard, A. 1974. *Génétique des populations humaines*. Paris: PUF.

Jacquard, A. 1977. *Concepts en génétique des populations*. Paris: Masson.

Jencks, C., et al. 1972. *Inequality: A Reassessment of the Effect of Family and Schooling in America*. New York: Basic Books.

Jensen, A. 1969. "How Can We Boost IQ and Scholastic Achievement?" *Harvard Educ. Rev.*, 39:1–123.

Jensen, A. 1974. "Kinship Correlations Reported by Sir Cyril Burt." *Behavioral Genetics*, 4:1–28.

Kamin, L. 1976. "Heredity, Intelligence, Politics and Psychology I." *The IQ Controversy*, pp. 242–264. New York: Pantheon.

Karlin, S. 1975. "General Two-Locus Selection Models: Some Objectives, Results and Interpretations." *Theor. Popul. Biology*, 7:364:398.

Kempthorne, O., E. Pollback, and T. Bailey. 1977. *Proceedings of the International Conference on Quantitative Genetics*. Ames: Iowa State University Press.

Kimura, M. and T. Ohta. 1971. *Theoretical Aspects of Population Genetics*. Princeton: Princeton University Press.

Langaney, A. 1977. "La Quadrature des races." *Génétique et Anthropologie, Science et Vie*, September, pp. 83–127.

Langaney, A. 1978. "Le Paradoxe du sexe et de la fortune," *Le Monde*, February 1.

Lefebvre-Witier, Ph. 1974. "Structure génétique des systèmes sanguins erythrocytaires et sériques chez les Kel Kummer." *Population*, 29:517–527.

Lerner, I. M. 1968. *Heredity, Evolution, and Society*. San Francisco: Freeman.

Leviandier, Th. 1975. "Loi de probabilité dunombre d'ancêtres et de descendants dans une population fermée." *Social Sc. Inform.*, 14:183–216.

Lewontin, R. 1974. *The Genetic Basis of Evolutionary Change*. New York: Columbia University Press.

Lewontin, R. 1976. "The Analysis of Variance and the Analysis of Causes." In *The IQ Controversy*, pp. 179–193. New York: Pantheon.

Loehlin, J., G. Lindzey, and J. Spuhler. 1975. *Race, Differences, and Intelligence*. San Francisco: Freeman.

Malécot, G. 1948. *Les Mathématiques de l'hérédité*. Paris: Masson.

Mendel, G. 1956. "Versuche über Pflanzen-Hybriden." In Gedda L., *Novant Anni delle Leggi Mendeliane*, pp. 3–100. Rome: *Instituto Gregorio Mendel*.

Montaigne. 1962. *Œuvres complètes*. Paris: Pléiade/Gallimard.

Morin, E. 1973. *Le Paradigme perdu: la nature humaine*. Paris: Édition du Seuil.

Morin, E. and M. Piattelli-Palmarini. 1974. "Ethique et science de l'homme" *L'Unité de l'homme*. pp 790–815 Paris: Édition du Seuil.

Morton, N. E. 1961. "Morbity of Children from Consanguineous Marriages." *Progress in Medical Genetics*, pp. 261–291. New York: Grune and Stratton.

Mourant, A. E. et al. 1976. *The Distribution of the Human Blood Groups and Other Polymorphisms*. London: Oxford University Press.

Nadot, R. and G. Vayssex. 1973. "Apparentement et identité: algorithme du calcul des coefficients d'identité." *Biometrics*, 29:347–359.

Neel, J. V. 1962. "Diabetes Mellitus: A 'Thrifty' Genotype Rendered Detrimental by 'Progress'?" *Am. Journ. Human Genet.*, 14:353.

Nei, M. 1966. *Molecular Population Genetics and Evolution*. Amsterdam: North-Holland.

Olivier, G. et al. 1977. "L'Accroissement de la stature en France. II. Les Causes du phénomène: analyse univariée." *Bull. et Mém. Soc. Anthrop.* Paris, 13:205–214.

Petit, C. and E. Suckerkandl. 1976. *Génétique des populations. Évolution moléculaire*. Paris: Hermann.

Prigogine, I. 1977. "L'Ordre par fluctuations et le système social." In A. Lichnerovicz and F. Perroux, *L'Idée de régulation dans les sciences*, pp. 153–191. Paris: Maloine-Doin.

Reed, T. and J. Neel. 1959. "Huntington's Chorea in Michigan, II. Selection and Mutation." *Am. Journ. Human Genet.*, 11:107–136.

Richard, J.-P. 1973. "Intelligence." *Encyclopaedia Universalis*, 8:1,081–1,084.

Robert, J. M. 1966. "La Génétique et la vie." Lyon: *Cah. Med. Lyonnais*.

Ropartz, C. 1971. "L'Allotypie des immunoglobulines humaines." *Bull. de l'Institut Pasteur*, 69:107–152.

Rosenthal, R. 1974. *On the Social Psychology of the Self-Fulfilling*

Prophecy: Further Evidence for Pygmalion Effects and Their Mediating Mechanisms, pp. 1–28. New York: MSS Modular Publication.

Rosenthal, R. and L. Jacobson. 1971. *Pygmalion à l'école*. Paris: Casterman.

Ruffie, J. 1976. *De la biologie à la culture*. Paris: Flammarion.

Russel, E. S. 1975. "Report of the Ad-Hoc Committee: Resolution on Genetics, Race and Intelligence." *Genetics*, 83:99–101.

Schull, W. J. and J. V. Neel. 1965. *The Effects of Inbreeding on Japanese Children*. New York: Harper and Row.

Segalen, M. and A. Jacquard. 1973. "Isolement sociologique et isolement génétique." *Population*, 28:551–570.

Sentis, Ph. 1970. "La Naissance de la génétique au début du XXᵉ siècle." *Cahiers d'Études Biologiques*, 8–19:73–85.

Stern, C. 1960. *Principles of Human Genetics*. San Francisco: Freeman.

Sutter, J. 1950. *L'Eugénique*. Paris: PUF-INED.

Sutter, J. 1968. "Fréquence de l'endogamie et ses facteurs au XIXᵉ siècle." *Population*, 23:303–324.

Sutter, J. and L. Tabah. 1951. "Effets des mariages consanguins sur la descendance." *Population*, 6:59–82.

Thuillier, P. 1974. "Les Scientifiques et le racisme." *La Recherche*, Paris, May.

Tobias, C. A. 1977. "Biological Effects of Radiation." *Encyclopaedia Britannica*, 15:378–392.

Verscheur, O. Von 1943. *Manuel d'eugénique et hérédité humaine*, Paris: Masson.

Vianson-Ponté, P. 1977. "Ce qu'on n'ose pas dire." *Le Monde*, June 18.

Wade, N. 1976. "IQ and Heredity: Suspicion of Fraud Beclouds Classic Experiment." *Science*, pp. 916–919.

INDEX